BREAKTHROUGH

HOW TO MAKE YOUR RESEARCH MATTER

DOUGAL EDWARDS

Typeset and printed by Drive Press

ISBN: 978-1-7642513-0-3 (hardcover)
978-1-7642513-1-0 (paperback)
978-1-7642513-2-7 (ebook)

NATIONAL LIBRARY OF AUSTRALIA

A catalogue record for this book is available from the National Library of Australia

Acknowledgements

There is a version of a researcher's career that appears in grant applications and as the introductions people give you at conferences. Then there is the version that only emerges in conversations, usually late in the day, when the professional caution is set aside and the truth comes through. I am grateful to every researcher who trusted me enough to share that second version. Each added something to an approach that only became useful because they were generous enough to test it in practice. The researchers you'll meet in this book are those who helped me see the furthest. And while I won't name them here and their details have been changed, they will recognise what they gave me. I hope they will also recognise that whatever is right about this book belongs to them far more than it belongs to me.

An approach only becomes reliable once enough people have helped you prove it somewhere real. Anita Hill, former Executive Director for Future Industries at CSIRO (Commonwealth Scientific and Industrial Research Organisation) and its first Chief Scientist, was the first to take that risk, and the confidence she extended at a moment when the approach had more ambition than evidence is something I have never forgotten. Steve Wesselingh opened the doors of SAHMRI (South Australian Health and Medical Research Institute) and gave me the kind of freedom that only arrives when a leader genuinely doesn't know what the answer is and is strong enough to admit it. Anne-Laure Puaux at WEHI (Walter and Eliza Hall Institute) and Julia Renaud at TRI (Translational Research Institute) brought me in and invited me back year after year, each return creating the conditions for refinement and progress that no single engagement could have produced. Tanisha Carino was directing FasterCures when she invited me in, and in doing so, accelerated everything. The foundation leaders and mission-led philanthropists I met through her changed the speed and ambition of what would become possible.

To Clare, Julia, Alex and Raul – thank you for everything you brought to our work together. To Kristen, you are the editor and publisher every writer hopes for and most never find – thank you.

To my mother Victoria, who raised me to be a better version of myself, my father Derek, who encouraged me to commit my career to the people who needed someone in their corner and my sisters Abigail, Gemma and Eleanor, who made me see the world from other angles – thank you.

Alessandra, you are the reason I can do work I believe in. Thank you for sharing this life with me. To my teenagers Sofia and Orlando, by the time you are fully in the world and able to appreciate the gap this book seeks to close, I hope it reads as history.

And to the researchers who have arrived at the question of how to get your work into the world and found the path less clear than promised, this book was written for you. Thank you for reading.

Praise for *Breakthrough*

This book grew from 15 years of working alongside researchers in laboratories, hospitals and universities across Australia. Here's what they've said.

'For a researcher of my vintage, "engaging industry" was such an abstract concept I had no idea what it really even meant. This opened a curtain to a larger landscape of science and discovery translation, commercialisation, and marketplace delivery.'
– Jenni Gunter, Queensland University of Technology

'It challenged the way I thought about my research and its potential value and impact, and provided a framework to transform what was in essence a "pie-in-the-sky" idea into a feasible project with tangible objectives and milestones.'
– Kristen Radford, Mater Research

'I have no doubt that I will benefit from all this for many years to come. It has already changed the way I think about how I might sustain a career in science in the long term. Given the exquisite fragility of the grant system at the moment, it feels good to have another bow in the quiver, and in years to come, I will describe this as a career game changer.'
– Aideen McInerney-Leo, University of Queensland

'The concept of an entrepreneurial researcher is right on target, and practical, concrete advice was provided that is helping me refine and extend my approach with industry partners.'
– **Michael Davern,** University of Melbourne

'It meant that we had to put our reflective glasses on, and think about why we do our research, who we're doing it for, what our findings mean, and ways in which to move the needle.'
– **Arutha Kulasinghe,** Wesley Research Institute

'It created a completely non-judgemental, deeply supportive environment where I could step outside the academic bubble and genuinely explore the real-world impact of my research. I learnt a huge amount and now feel equipped and excited to put these skills into practice.'
– **Sara Berent,** Walter and Eliza Hall Institute

'It helped frame what is important at the current stage of my start-up. It's overwhelming to think about everything that needs to be done, but having someone explain that it's normal to feel like that, that what you have is good and you only need these specific things to make it great, gives you the confidence to prioritise.'
– **Andrew Peska,** Peter MacCallum Cancer Centre

'There is a world of untapped potential for their ideas and science beyond what they think of during an academic career. Even if they don't decide to venture on a commercial path, the skills and tools will make their research, and ability to present it, vastly better.'
– **Niall Geoghegan,** Walter and Eliza Hall Institute

'We came with an exciting new technology but none of the communication or commercialisation skills that could supercharge our idea. In hindsight, this upskilling was the critical first step in building our company. Two years on, our journey has led to us successfully launching Proxima Bio as co-founders.'
– Jon Bernardini and Jason Brouwer, Proxima Bio

'Support went far beyond advisory. They helped us not only refine our strategy but also navigate complex stakeholder dynamics with clarity and confidence. I highly recommend this to any venture looking to bridge the gap between technical feasibility and commercial success.'
– Natalie Thorne, Genomical

Table of contents

INTRODUCTION

In 2015, in front of a full-to-capacity conference of early and mid-career researchers, two worldviews collided.

The theatre was packed with PhDs and postdocs who had organised a symposium for their peers at a national research organisation. Between internal project briefings, external speakers had been invited to offer perspectives from beyond the research world. I was one of them. But I wasn't there just to offer a new perspective. An executive director had brought me in to run a pilot accelerator programme for deep technologies, and I was there to talk about what we were attempting to do and to find researchers willing to join us.

This accelerator programme had real stakes. Over the life of the programme, researchers would be required to test their research against real market problems, speak directly with potential users and customers and demonstrate that their work could translate into something people actually wanted or needed. Projects that couldn't find a market would be killed. But those that the researchers could validate would become the foundation for actual ventures. This was a test of whether researchers could learn to connect their work to real demand.

1

This is what I was prepared to share at the conference that day – a presentation that painted a picture of what might be possible. A future where researchers didn't have to hand their work off and just hope it made it to market, but could guide it into the world themselves – on their terms. A future where agency wasn't something granted by your position in the hierarchy, but was something you claimed through action.

The room was engaged, and I could feel the message landing. It was giving a language to something many of the audience had already felt: a frustration with being a passenger in someone else's career development, a hunger for meaningful work and a feeling that there had to be more to research than publishing papers and hoping someone noticed.

Then a hand went up near the front of the audience. I recognised the woman immediately. She was a senior research leader at the organisation holding the conference, and she was both respected and accomplished by those in her industry. But, importantly, she had built a successful career entirely within the traditional research system. We'd met before, and in that fraction of a second before she spoke, I allowed myself to hope she might be about to back me up, and give this thing some institutional weight.

The room quietened as she began speaking.

'These people should be prioritising publications,' she said. 'Not this stuff.'

I felt my heart sink. I could almost see the programme withering in front of me, imagining researchers walking away and the pilot dead before it even got started. I thought about all the work the executive director and I had put in to get this over the line – it was about to be an experiment that would never run because a respected voice had told the room it wasn't worth their time.

I responded the only way I could, by pointing to what was already happening elsewhere. The National Science Foundation (NSF) in the US had been running programmes like this in pockets, and the UK was just starting too. The evidence was public if you knew where to look, and the early results were promising. This wasn't a theory about what might work but a model proving itself for researchers around the world. It was, as I saw it, the future.

But in reality, her challenge did everyone in that room a favour, because it forced a fault line to open that had been invisible until someone said it out loud. Before she spoke, two worldviews had coexisted almost unnoticeably. Researchers could feel that something had been changing around them. Things were getting harder to get across the line, from research projects to programmes, funding was more challenging and funders the world over were seeking out proposals that were closer to the finish line. But senior figures in universities and institutes continued to state exactly what this leader had just said, dispensing advice that had worked for them in their careers without asking or even wondering whether that path still existed.

Her challenge ended that silent coexistence, because she forced people to see and defend, out loud, the assumptions the traditional system quietly reinforced – that papers and publications come first, impact comes later and any talk of agency is a divergence from the status quo.

Half the room wanted to believe her, because if she was right then the path forward was clear and established. They knew what to do. Publish, apply, wait your turn, trust the system. But the other half had recognised it for what it was. The white-knuckled grip on a system that wasn't responding to modern pressures and needs. And now that these two systems were clearly defined, they had a choice. They could return to the congested pathway they knew, or they could test whether another, better, route existed.

I finished my talk, expecting to pack up quietly and slip out the back. Instead, a queue formed next to me. Researchers at every level – doctoral students, postdocs and group leaders – all wanted to talk. These weren't the usual networking conversations, punctuated by the polite exchange of business cards and vague promises to connect. These were real and specific questions. How do I learn more? What might this look like for my work? And then invitations: come present to my supervisor, talk to my team, see what we're working on and tell us if any of it might work. In the following weeks, the applications arrived. The programme was alive.

That pilot would go on to seed a national programme that still runs today, one that has helped thousands of researchers find a path for their science into the world. The new system was already off and running, and now we were starting to be ready for it.

The system's failures aren't your failures

Before we go any further, it's worth naming something that you might be carrying with you, consciously or unconsciously.

Researchers regularly talk about their heavy workloads and the legitimate busy life of science. They complain while passing in the hall, or grabbing a coffee, that grants are harder to get and worth less when you get them. They worry about the publishing treadmill that everyone's on, noting that few seem to be really getting anywhere. Everyone in the tea room nods along with these conversations because everyone in the tea room feels and sees it too. They're having the same experience. And then the conversation moves on because this kind of complaint is allowed. It's structural. It's universal. It's not about *you*.

But sometimes, usually late at night or after a few drinks or in the moments of honesty that come when someone's had a particularly brutal rejection, something more starts to surface. A suspicion that maybe the problem isn't the system at all. Maybe it's *me*. Maybe I'm just not as good as the people who seem to be making it work. Maybe if I were sharper, or more strategic or just worked harder, I wouldn't be struggling the way I am.

This is the part that doesn't get said in the tea room, because once you say it out loud you can't unsay it, and because the fear that you're not enough is the kind of thing that burrows in and starts to feel like truth.

And beneath even that is the place researchers almost never go because going there calls everything into question – the job, the work, everything that researchers have bought into throughout their entire career. And it's here that we find the real fear.

What if the path itself is broken?

What if it's not that I'm failing to walk it properly, but that it doesn't lead anywhere at all anymore?

What if all this effort, all these years, all these sacrifices, are carrying me down a road that used to reach somewhere but now just goes on and on until I simply run out of energy to keep moving forward?

If any of that sounds familiar, if you've felt that creeping suspicion that maybe the path itself has taken a wrong turn somewhere, it's not you, and it never was.

The system you trained for was built for a world that no longer exists. That's because the infrastructure that once connected discovery to deployment, the whole elaborate bridge between academia and the world, has all but been dismantled piece by piece over the last 30 years. What remains is a system that trains researchers for one world and then drops them into another entirely different world, and in the process continues to measure their performance against criteria that no longer predict anything meaningful.

The system's failures are not your failures, and there is another way through.

The fork in the road

That day in the auditorium was a fork in the road, not just for the researchers who joined our pilot, but for me as well.

I hadn't come from research. I'd spent the previous years building startups, advising founders and helping industry figure out their approach to venturing and innovation. This was work built around a discipline the research world had never been forced to learn: *you don't know what you have until someone outside your world tells you what it's worth*. The world provides you that feedback whether you want it or not, and this feedback is important information about what the market wants and needs, so long as you're willing to listen. The teams I worked with either found traction or they failed, but in both situations, they found out within months, rather than wasting years on work that wasn't viable.

But I wasn't yet in the research industry, so the reason I was even in the room delivering at that symposium in 2015 was down to a chance discussion a year before. An executive director at that organisation had been watching the work that startups were doing. She had noticed that though her scientists were producing excellent research, too little of it was reaching the world. The gap between discovery and deployment kept widening, and nothing in the traditional system seemed able to close it. She asked me whether the methods that worked for my startups and industry might work for researchers too.

From her own admission, the question had seemed almost naive at first. Researchers operate under different logics, different timelines and entirely different definitions of success. What could methods built for a world driven by a 'move fast and break things' mentality possibly offer the world of research? However, it was anything but naive.

The more we talked, common problems came into focus. Startup founders were still building things without knowing whether anyone needed them. Industry still added product features while assuming customers would pay. And researchers still optimised for publications while omitting to ask whether anyone would use the work. Each group was measuring itself by what was easy to measure, not what really mattered.

What did separate these worlds, however, was the outcome that arose from this lack of feedback. Industry projects and ventures died quickly, usually within months, when they got it wrong, because the market told them. But researchers could spend their entire careers navigating the research system without ever discovering that their work was useless. Affectionately referred to as 'zombie projects', these were the bane of leaders across institutes, as they silently sapped up valuable resources and talent, without ever having a real chance to succeed.

We decided that the work we were doing with startups and industry could help solve that problem. It could help researchers find out much earlier if their work was worth pursuing.

The executive director was firm. 'I believe that those methods will benefit my researchers,' she asserted. 'Let's find out.'

That pilot programme was our experiment together. The collision in the auditorium with the senior researcher was its first public test. And the queue of researchers afterwards told me something I hadn't expected: the hunger was real, even if most of them couldn't yet name what they were hungry for.

Where this book came from

What I didn't know, standing in that auditorium, was that the next decade of my life had just taken shape. I'd go on to work with hundreds of researchers navigating this systemic shift, learning to test their science against real demand, finding pathways the traditional system never signposted and raising the capital to translate their work on their own terms. I deliberately worked the other side too, advising foundations and philanthropists searching for researchers ready to work differently but who were struggling to find them inside the traditional system.

Through that decade, we proved that despite the lack of infrastructure connecting researchers to impact, there were and are systematic ways to increase your success, and many researchers found partners, funding and paths to take their work into the world. Of course, others discovered their science wasn't ready, or that the market they'd imagined didn't exist, but they discovered it quickly, before the best years of their careers disappeared into the gap. Both of these results were progress, and with each researcher through the pilot programme, our approach sharpened and we gained more information, more understanding and more traction.

This book distils what that decade revealed.

Basic research still matters

The senior research leader who stood up in that auditorium was protecting something real. Because basic research matters. Basic research produces science that shapes fields across the globe. And the traditional system enables that. The rigour it instilled, the depth it rewarded, the discoveries it enabled – none of that is illusion. Deep, curiosity-driven inquiry remains the foundation everything else is built on. And publishing does matter.

Moreover, the advice she was giving was right for the world she'd come up in. During most of her career, the downstream infrastructure did exist, because industry laboratories caught promising science and carried it forward. You could prioritise publications, build your reputation and trust that quality would eventually be recognised by the world. That path worked because the bridge the old system had built between discovery and deployment was still standing, and the traffic across it was light enough that the gatekeepers were happy to help research across.

What I was trying to articulate at that symposium, however imperfectly, was that the bridge had already started to collapse. The infrastructure connecting basic research to the world has all but been dismantled. And no amount of excellent science will rebuild it.

The funders who recognised this dismantling first were disease foundations, patient advocacy groups and philanthropists, but that doesn't mean they've stopped valuing basic research either. Instead, the more progressive ones stopped waiting for it to cross a bridge that no longer existed. And they started building new bridges, ones without gatekeepers, ones they were in control of, and they started searching for researchers willing to make the crossing.

Agency over permission

The researchers I've worked with who've made this crossing success-fully are still researchers. They still care about the science. They still do the work properly. They just stopped waiting for a system that was no longer able to help them. These are the researchers who decided to act.

What distinguishes them isn't that they're more entrepreneurial by nature, or more willing to compromise or less committed to doing things right. It's simpler than that. At some point, each of them recognised that the landscape had changed, and they decided to change with it.

Some were pushed into action, because funding dried up or a career path closed or circumstances left them no choice. Others were pulled to act because they saw something emerging and wanted to be part of it. They all felt the same fears and doubts that any researcher feels when they contemplate stepping off the path they were trained for. But they took that step anyway.

What they found was a new path that was able to provide what they had always been promised by the traditional system, but it had stopped being able to deliver. They found funding that didn't require years of grant applications. They found partners who accelerated their work instead of slowing it down. And, most importantly, they got to repeatedly enjoy the genuinely wonderful experience of watching their research reach the people it was meant to help.

The path they found isn't hidden. The funders aren't gatekeeping it, and the capital isn't scarce. What's missing is researchers ready to step onto it.

Are you ready to cross a new bridge?

Two systems now exist for carrying research towards impact, and each has built its own bridge across the valley that separates discovery from the world.

The old system is the one you were trained for. Its stakeholders built a bridge decades ago, when corporate R&D labs and pharmaceutical companies were actively pulling promising science across. That bridge still stands, though it's crumbling and overcrowded, and the gatekeepers who control it have learned to benefit from the scarcity. The tolls are high. The requirements are opaque and regularly shift. You petition, and they decide. Even researchers who pay every toll and navigate every checkpoint often arrive to find the terms have changed, the opportunity has narrowed, the value they could create has flowed somewhere else.

The new system was built by different stakeholders: disease foundations and venture philanthropists and patient advocacy groups who got tired of watching excellent research disappear into journals while patients waited and died. They've built a different kind of bridge, one where the currency isn't permission but demonstrated value. The requirements are legible. The milestones are clear. If you've done the work to show your work matters, you're welcomed with open arms, because the people who built this bridge succeed only when you do. Their incentives align with yours in ways the old system's never did.

That same choice is in front of you now. The shift is still unfolding and the new bridge is still being built and the funders are still searching for researchers ready to cross it. This all means you are not too late. In fact, if you act now, you're still early.

In this book

First, you need to see the old system clearly, because until you can, you'll keep mistaking its failures for your own. That's where we'll start, with the story of how we found ourselves in this situation, not because history is interesting for its own sake but because understanding what happened is the first step to knowing where you now stand. And once you can see that the bridge you were trained for is overcrowded and failing, that it's this that keeps you and your research in the doldrums, and that there's the possibility of another way, that feeling of helplessness turns into one of hope, and the levers for change become visible.

From there, we'll move to exploring how to navigate and cross the new bridge. We'll address the practical question of how you build the demonstrated value that earns the crossing, how you read the forces shaping your work and test your assumptions and how you can build traction without waiting for anyone's permission.

And finally, we'll talk about what it means to lead from wherever you happen to stand, not through rank or title but through action, through becoming the kind of researcher the new system is searching for and the old system never taught you to be.

The shift is real, the funders have moved and the researchers who see both bridges clearly will have choices the others won't.

The only question left is whether you're ready to cross.

PART I

DISCOVER

The traditional system that trained you only shows you one bridge – the one your supervisors crossed, the one the metrics continue to point towards, the one that feels like the only legitimate path because everyone you trust has told you it is.

That bridge still stands, technically, though it's overcrowded and crumbling and controlled by gatekeepers who benefit from the scarcity that it creates (whereas you, as the researcher, do not). From inside that system, it still appears that nothing fundamental has changed, that the path your supervisors walked is the same path available to you, it's just more competitive now and requires more effort.

Part I is about seeing this clearly for the first time.

By the end, you'll understand why the bridge you were trained for can no longer carry its load, and you'll see that another bridge exists, built by people who are actively looking for work like yours. You just have to take the first steps across.

Chapter 1
MIND THE GAP

Sarah had spent nearly two decades as a clinician, sitting across from patients who needed answers she couldn't always give them. Eventually, she decided that if the science didn't exist to close those gaps, she'd build it herself, and so she left clinical practice for research.

She followed her research career across three continents, published in respected journals and built the kind of track record that opens every door the system has to offer. When she won a research fellowship to pursue research that could bridge the gap between what her science could now do and what patients actually received, it felt like the beginning of everything she'd spent her career working toward. But within a year, she understood that the system she'd entered had no interest in helping her close that gap, and no mechanism for doing so even if it had.

On paper, she had arrived. But in practice, she was stuck — not at the starting line where at least you know which direction to run, but somewhere crueller. She was stuck on a plateau. She'd done everything she was supposed to do — built the thing she was supposed to build, reached the place she was supposed to reach and yet there she stood, with no way to bring this work to the people who really needed it.

If she'd been failing, at least she'd have known what to fix. Failure is clarifying in its way, because you can look at what went wrong, adjust and try again. But she wasn't failing, and that's what made it so hard to talk about. The research was ongoing. The publications were happening. The grants were still coming through, at least some of them. Her lab was functioning. The work just wasn't landing where she needed it to. It wasn't creating the traction or the partnerships or the momentum that would carry it from the lab into clinics and lives.

She'd find herself in meetings with industry partners where the dynamic felt off. She knew what her work could do for patients, but those partners couldn't see past the academic framing to recognise it. She'd watch her research generate findings that could genuinely help patients, and then watch those findings sit in journals where, in a kind of incestuous cycle, the only people reading them were other researchers for the purpose of citing them in their own papers that would then just sit in other journals, waiting for their own citations.

Sarah found herself wondering, in the quiet moments between grant deadlines and lab meetings, what success would even look like for her now. It certainly wasn't success as the system defined it, because she had that. But the kind of success she'd imagined when she first decided to become a researcher. The kind where your work actually reached the people it was meant to help. She began to worry that the work she'd come into research to do, work she'd chosen precisely because she'd seen what patients needed, might never actually reach them.

Let's go back to the maths

The mathematics that explain Sarah's plateau are worth understanding. Firstly, because once you see them you can't unsee them and, secondly, because they explain so much of what researchers feel but struggle to name.

Between 1950 and 2016, the US's real GDP increased by a factor of about eight, which represents substantial economic growth by any measure. But during that same period, the research industry grew far faster. The number of PhDs granted grew more than sixteenfold. The National Institutes of Health (NIH) and NSF funding increased 75 times in real terms. And the number of publications rose more than twentyfold.[1]

By every measure, science expanded faster than the economy that was supposed to absorb its outputs. And it hasn't stopped. Around 90% of all researchers who have ever lived are alive today – a number I've had to check several times because it seems too dramatic to be true.[2] In fact, every year the world produces more PhDs than the entire global university system can absorb.[3]

So the research industry has grown and the funding pie hasn't kept pace. In real terms, research funding has remained relatively flat for

---

1   Collison, P & Nielsen, M. (2018). 'Science Is Getting Less Bang for Its Buck.' *The Atlantic.* Retrieved March 6, 2026 at https://www.theatlantic.com/science/archive/2018/11/ diminishing-returns-science/575665/.
2   UNESCO. (2021). UNESCO science report: The race against time for smarter development. UNESCO. Retrieved March 11, 2026 at https://www.unesco.org/reports/science/2021/en.
3   Matsoukas, I. (2025). 'Prestige without purpose? It's time to reinvent doctoral education.' Times Higher Education. Retrieved March 6, 2026 at https://www.timeshighereducation. com/blog/prestige-without-purpose-its-time-reinvent-doctoral-education.

the last 30 years across most of the developed world.[4] There was a boom in the second half of the 20th century, an extraordinary expansion that created the modern research enterprise as we know it, and then the boom ended, while the system kept running as if it hadn't.

The result is what you'd expect: fierce competition for shrinking returns. Researchers today spend a third to half of their time preparing and applying for grants.[5] Preparing a single major proposal can consume two months of concentrated effort – two months of early mornings and late nights, of drafting and redrafting, of pulling together letters of support and budget justifications and impact statements, two months when the research itself has to wait while you write about the research you hope to do. And at current success rates, it takes the equivalent of 300 working days of effort to get one proposal funded.[6]

In Europe, European Research Council success rates hover around 11 to 14%.[7] In the US, NIH success rates have dropped from over 30% in the early 1990s to roughly 17 to 20% today.[8] The pattern holds everywhere you look, across Australia and Canada and the rest of the

---

4   Innovative Research Universities. (2023). 'Major trends in university research income and expenditure.' Retrieved March 6, 2026 at https://iru.edu.au/wp-content/uploads/2023/03/Major-trends-in-university-research-income-and-expenditure.pdf.

5   Lofgren, E. (2021). 'Top researchers spend 50% of their time writing grants, how to fix it, and what it means for DoD.' Acquisition Talk. Retrieved March 6, 2026 at https://acquisitiontalk.com/2021/12/top-researchers-spend-50-of-their-time-writing-grants-how-to-fix-it-and-what-it-means-for-dod/.

6   Schweiger, G. (2023). 'Can't We Do Better? A cost-benefit analysis of proposal writing in a competitive funding environment.' PLOS ONE.

7   European Research Council. (2023). 'ERC Starting Grants 2023: Statistics.' Retrieved March 6 2026 at https://erc.europa.eu/sites/default/files/2023-09/erc-2023-stg-statistics.pdf.

8   DrugMonkey. (2025). 'Fun with historical NIH funding success rates.' Retrieved March 6 2026 at https://drugmonkey.wordpress.com/2025/07/30/fun-with-historical-nih-funding-success-rates/.

developed world, and the direction is always the same.[9]

At least a third of the research time of our brightest minds is disappearing into applications that will come to nothing – not because the ideas aren't good or the researchers aren't talented, but because the mathematics of the system have become impossible.

This alone would explain why so many researchers feel stuck. But competition is only part of the story, and not the most troubling part.

Here's what makes the situation genuinely concerning.

We're not pushing the innovation frontier

If the problem were simply that more researchers are competing for the same resources, you'd expect the system to be producing more breakthroughs than ever. More minds working on more problems with more sophisticated tools should, in theory, generate more discoveries. In most situations, more competition drives more innovation. So, the sheer volume of talent and effort pouring into research should be pushing the frontier forward faster than at any point in history.

It hasn't.

Despite all those inputs, including more researchers than ever, more publications than ever and more sophisticated tools than any previous

---

9   Australian Research Council. (2024). 'Discovery Projects Selection Report for funding commencing in 2024.' Retrieved March 6, 2026 at https://www.arc.gov.au/discovery-projects-selection-report-funding-commencing-2024; Canadian Institutes of Health Research. (2012). 'Evaluation of the Open Operating Grant Program – Final Report 2012. Retrieved March 6, 2026 at https://cihr-irsc.gc.ca/e/45846.html.

generation could imagine, the rate of transformative discovery has declined. We're producing more science than at any point in human history, yet the science we're producing is, on average, less significant than what came before.

Economists studying research productivity have found that ideas are getting harder to find, and the effort required to generate a given amount of innovation has increased dramatically over time.[10] In semiconductors, the number of researchers required to double chip density has increased eighteenfold since the 1970s. In agriculture, the effort required to maintain historical yield improvements has grown fourfold. In medicine, the number of researchers needed per new approved drug has increased fivefold.[11]

More effort is producing less transformation, and the pattern holds across virtually every field of research we have examined.

At the level of individual papers, the trend is equally clear. A comprehensive study of 60 million research articles found that work is increasingly likely to consolidate existing knowledge rather than push into new territory.[12] The share of papers that disrupt or redirect a field, rather than merely extend it, has fallen steadily across every scientific domain since 1945. Science is becoming more incremental, more derivative and more likely to add a brick to an existing wall than to lay a new foundation.

But now we need to ask ourselves… why?

10 Bloom, N, Jones, C, Van Reenen, J & Webb, M. (2020). 'Are Ideas Getting Harder to Find?' *American Economic Review*. Retrieved March 6, 2026 at https://www.aeaweb.org/articles?id=10.1257/aer.20180338
11 Bloom. 'Are Ideas Getting Harder to Find?'
12 Park, M, Leahey, E & Funk, R. (2023). 'Papers and patents are becoming less disruptive over time.' *Nature*.

What's really going on?

There's an obvious explanation for this, and it's partly true. The frontier has moved. The easy discoveries have been made. So now, each step forward requires standing on an ever-increasing pile of previous knowledge, and the distance that has to be covered before you can meaningfully contribute grows with every iteration.[13]

But that same study of 60 million papers pointed toward a different explanation. And this one has less to do with the inherent difficulty of discovery and more to do with what the system rewards.

The reality is that researchers producing incremental work today aren't less talented than their predecessors. They're doing what makes sense in a system that has turned inward, optimising for what it can measure and control. Grant committees are faced with far more proposals than they can possibly fund, so they need reasons to say no. A proposal that promises something ambitious is a liability, because ambitious means risky, and risky means it might not deliver. These are the easiest to put on the chopping block. But a proposal that guarantees modest but certain outputs is safer for everyone involved. And these tend to get through.

The committee members aren't being malicious; they're being rational given the metrics they're measured by and the constraints they face. But the cumulative effect of this systemised approach, across thousands of funding decisions year after year, is that bold proposals continue to lose out to cautious ones over and over again.

---

13 Jones, B. (2005). 'The Burden of Knowledge and the "Death of the Renaissance Man": Is Innovation Getting Harder?' *The Review of Economic Studies*.

The same dynamic operates in journals. Editors prefer papers that extend existing work because they're easier to evaluate, more likely to be cited and less likely to be wrong in embarrassing ways. Researchers, watching all of this, adapt accordingly. They split one definitive study into four smaller publications because they recognise this will be easier to get across the line. They propose safe bets even when they know, when they're being brutally honest with themselves, that this isn't the most important work they could be doing.

Researchers who play the game, who put forward modest proposals and incremental papers, aren't cynics or sellouts. They're survivors, and they're just doing exactly what the system teaches them to.

Sarah felt all of this without having the language for it. She had entered research under a social contract she'd never seen, one drafted at a time when almost none of what she took for granted existed yet, when universities were teaching institutions and corporate laboratories barely existed and the whole apparatus of government-funded research was being invented from scratch.[14]

Nobody had shown her that contract or explained its terms, but she'd absorbed them completely – she knew that governments fund discovery, researchers produce knowledge and then someone downstream carries it forward. That deal had worked for her supervisors and for their supervisors before them, and she'd been told that it would work for her too. But nobody had mentioned that the downstream infrastructure – the corporate laboratories and development pipelines that once caught promising science and carried it toward application

---

14 Here we are referring to the system posited by Vannevar Bush. When he got approval to move forward with his thesis post WW2, he laid the formal foundations of the old system – and then the infrastructure built itself around and on top. We will discuss this in greater detail in the next chapter.

– was being quietly dismantled while she was being trained as if it all still existed.

Sarah's experience isn't an anomaly. It's the predictable outcome of how the system now works. The bridge between discovery and deployment in the traditional researcher system has collapsed. What remains is a system that prepares researchers for one reality and drops them into another, then assesses their performance against criteria that no longer predict anything meaningful for them or their work.

Researchers are paying the price

The toll of all this shows up in ways the system doesn't measure.

Rates of anxiety and depression among researchers exceed those of the general population, and the gap has been widening.[15] This is what you'd expect when an entire profession is caught between what they were trained to do and what the world now asks of them. The constant pressure to produce, combined with the risk-adverse filtering of grant outcomes and publications, creates conditions that erode wellbeing in ways that are hard to see from the outside but impossible to ignore from within.

So people are leaving. Quietly, without fanfare, talented researchers are walking away from their chosen field, not because research is hard, since it's always been hard and they knew that going in, but because the relationship between effort and meaning has broken down. And nobody can explain why. Nobody can explain why the unspoken

---

15 Evans, T, Bira, L, Gastelum, J, Weiss, L & Vanderford, N. (2018). 'Evidence for a mental health crisis in graduate education.' *Nature Biotechnology*.

contract isn't delivering, let alone what they can do about it. So they make the only conclusion they can. They decide the problem must be them. If they were just smarter or more strategic, they'd be thriving.

But it isn't them. It was never them. And it isn't you, either.

Seeing the shift

When Sarah finally understood all of this, that it wasn't her that was broken but the system she was operating in, something shifted.

The relief was almost physical. Not because the situation was good (it wasn't… yet), but because it wasn't her fault. The years of wondering whether she was smart enough, strategic enough, productive enough, or whether the problem was some deficiency in her that she couldn't quite name – all of that had been misplaced. The system had changed around her while she kept following the rules she'd learned from the beginning. She'd been holding up her end of a bargain, but the system hadn't.

But just because she felt relief didn't mean she knew what to do. Relief isn't the same as resolution, and understanding why you're stuck doesn't unstick you.

Sarah had one more question to resolve, once the relief settled and the mathematics were clear. If the contract had worked for her supervisors, and for their supervisors before them, what had changed?

She knew that researchers hadn't become less talented. She knew science hadn't become less important. So something in the architecture itself must have shifted, some load-bearing pillar or pier had been removed, and the whole structure was now breaking in ways nobody had planned for.

To find her way, she would need to understand what had been built, and why it had stopped working.

Which meant going back to the beginning of modern research. Which is where we need to go as well.

Chapter 2

THE BRIDGE THAT DISAPPEARED

The pilot programme that started my decade-long quest to understand how we could systematically improve research's impact ran its course and exceeded even our best expectations. A national programme followed, and I found myself having conversations I wouldn't have had a year earlier, with people who saw the research system from angles I'd never encountered.

One of those conversations has stayed with me.

The boardroom belonged to one of Australia's most prominent investment firms, the kind of venture capital operation that positioned itself between university discoveries and commercial markets, backing research through the path to application. I'd knocked on this door a year earlier and had a polite conversation that went nowhere. Now the managing partner wanted to talk.

The conversation started cordially enough. We exchanged the usual pleasantries, in our case memories from our time as students in Europe. And then the conversation turned to discovery and whether or not we had any aligned incentives. It was the kind of exploratory discussion where everyone is trying to figure out whether there's something worth doing together.

Eventually, he asked what I was trying to accomplish working as I was with researchers. So I took the opportunity to explain my belief that researchers should have more agency over where their work goes and that the system would work better if the people who understood the science also understood how to connect it with the world.

He listened carefully, nodding in the places you'd expect and asking clarifying questions. Then he leaned back in his chair.

'You could help us by bringing us deals,' he said. *Great*, I thought! 'But not by upskilling researchers. We don't want them to think they can do this themselves.'

My heart sank. I asked why.

'It makes deals difficult,' he said. 'Researchers like these come in with unrealistic expectations. They want stakes that don't reflect the risk we're taking. They want to keep doing the interesting science instead of the experiments that would actually de-risk the invest-ment. They've read enough to be dangerous but not enough to be realistic.'

There was nothing hostile in it, and nothing particularly guarded either. Just the weary patience of someone who had seen this play out many times and had long since stopped questioning whether market-focused researchers made his job harder.

'The best deals,' he said, 'come from researchers who trust us to know what we're doing, who let us structure things properly, who focus on the science and let us focus on the business.'

What I heard beneath the frustration was something he perhaps hadn't fully articulated even to himself: that his firm's model worked best when researchers arrived knowing everything about their science and nothing about what it was worth. But when they understood markets and deal structures and comparable transactions, they became harder to work with, not because they were wrong, but because they had leverage his firm had grown accustomed to holding alone.

I asked what would actually help, from his perspective. If upskilling researchers made deals harder, what would make my work useful to someone in his position?

'Bring us deals,' he said again. 'We're always short of good deals. Researchers treat us like sharks in the water. They've heard the horror stories, so they're suspicious. They'd rather let their work rot in a journal than talk to someone like me. If you can get them comfortable enough to have the conversation, that's valuable. We'll take it from there.'

And if I couldn't bring deals?

He paused, and what he said next surprised me.

'Bring us competition.'

I asked what he meant.

'The market's too thin,' he said. 'When we're the only people in the room, researchers have no alternatives. Half of them won't engage at all because they think they're going to get screwed. If there were other venture funds, more researchers would start the journey. Some of them would end up with us anyway. The pie would get bigger.'

He wasn't being self-serving (or not *only* self-serving). He was describing the world as he found it. But he was also describing the endpoint of something that had been building for 70 years. A system designed with the best intentions that had slowly, invisibly, created the conditions for its own exploitation.

To see how we arrived at that boardroom, you have to go back to 1945.

1945 – the creation of the invisible contract

In July of 1945, Vannevar Bush faced an unusual problem.

For five years, Bush had directed the US Office of Scientific Research and Development, coordinating the Manhattan Project, the mass production of penicillin and the development of radar. He had watched science change the course of history in real time and had seen what happened when the full resources of a nation were directed toward solving problems that urgently mattered.

But now the war was ending. The urgency was fading. The wartime funding that had sustained thousands of researchers was about to disappear, and Bush understood that without a new justification, a peacetime rationale for continued public investment, science would retreat to the margins of national life. The momentum that had produced the atomic bomb and saved countless lives with antibiotics would dissipate, and the researchers who had made those achievements possible would scatter back to underfunded university laboratories or leave the field entirely.

The question Bush was wrestling with wasn't just his own. It had become a problem for the nation. And President Roosevelt wanted to know what America should do with its scientists now that the war no longer needed them.

Bush's answer came in a 40-page report called *Science: The Endless Frontier*, and its central argument was beguilingly simple.[16] Basic research, the kind of curiosity-driven inquiry that had no immediate application and might never have one, was nonetheless the wellspring from which all practical benefits eventually flowed. The scientists investigating fundamental questions about nature weren't wasting time on abstractions. They were filling a reservoir that applied researchers and engineers and industry would draw from for decades to come.

The model Bush proposed was linear, a kind of assembly line for knowledge:

Figure: The linear view of translation espoused by Vannevar Bush in 1945.

Each stage would feed the next in an orderly progression. Researchers would explore the unknown, driven by curiosity and the questions that logically followed. Engineers and applied scientists would take what the researchers discovered and figure out how to use it. Industry would develop those applications into products and services. And, ultimately,

---

16 Bush, V. (1945). 'Science: The Endless Frontier.' National Science Foundation. Retrieved February 10, 2026 at https://nsf-gov-resources.nsf.gov/2023-04/EndlessFrontier75th_w.pdf.

citizens would benefit. Bush believed that governments should continue to fund the science and trust the scientists, and, eventually, societal benefit would follow as naturally as water flows downhill.

Bush assumed that the transitions between the stages of the linear model would happen organically, just as atomic theory had given him the atom bomb. They'd be carried forward by those whose purpose was to catch what came before and carry it toward what came next. And this meant that the only job researchers needed to worry about was discovery. Translation was someone else's problem.

It was a beautiful theory, and it was also a political masterstroke, because Bush had found a way to justify public funding for curiosity-driven research by promising practical returns without requiring researchers to think about those returns themselves.

This was the contract, even if no one called it that at the time. And it didn't need to be written down because the infrastructure that made it work already existed and was so obviously aligned with commercial and societal benefit that no one imagined it could disappear.

Bush's report was American, but its influence spread across the developed world within a decade.[17] The UK established its own system of research councils. Germany rebuilt its Max Planck Institutes from the ashes of the war. France created its national research organisation, the Centre National de la Recherche Scientifique. Japan, Canada, Australia, the Netherlands, Sweden and dozens of other countries also constructed their own versions of the same basic architecture.

---

17 SRAI News. (2020). 'Science's 75 Year Pursuit of the Endless Frontier.' Society of Research Administrators International. Retrieved February 10, 2026 at https://www.srainternational. org/blogs/srai-news/2020/07/09/sciences-75-year-pursuit-of-the-endless-frontier.

Governments would fund basic research. Universities would conduct it. Industry would translate it. The public would benefit. The Bush model, exported and adapted, had become the operating system for research across the developed world.

And for decades, the model appeared to deliver exactly what Bush had promised. Research during this time led to the structure of DNA, the transistor, the polio vaccine, the laser and the Green Revolution, which transformed global agriculture and fed billions. These discoveries cascaded into application with enough regularity that the linear model seemed not merely plausible, but proven.

But the contract also contained an assumption so fundamental that no one thought to examine it, even though it was also fundamentally flawed – that the transitions between stages would always happen because someone would always be there to catch what researchers produced and carry it forward.

The linear model in operation

If you wanted to see that bridge in full operation in the 1980s, you could have visited a sprawling campus in Murray Hill, New Jersey, where a telephone company had built something that looked more like a university than a corporate facility.

Bell Labs, in its prime, employed 15,000 people, including 1,200 with doctoral degrees.[18] Eleven Nobel Prizes would eventually go to people who had done their breakthrough work within those walls, which

---

18 Gertner, J. (2012). *The Idea Factory: Bell Labs and the Great Age of American Innovation*. Penguin Press.

gives you a sense of the concentration of talent and the significance of what it produced. This earned Bell Labs the Guinness World Record for the most Nobel Science Prize winners from one laboratory.[19]

The researchers who worked there weren't trying to build better telephones. In fact, they weren't trying to *build* anything really, or at least not directly. Instead, they were investigating the fundamental properties of matter and energy and information, pursuing questions that might take decades to yield practical applications, if they ever yielded them at all. But applications did come.

The transistor emerged from researchers who were trying to understand how electrons moved through semiconductor materials, not researchers who were trying to invent anything in particular. Information theory, the mathematical framework that would eventually underpin everything from mobile phones to the internet, was developed by a mathematician named Claude Shannon who was simply trying to understand the fundamental properties of communication.

No one at the outset knew that work would become the foundation of the modern digital world. AT&T, who founded Bell Labs, funded all of this not as charity or to better their public relations, but as corporate strategy. And they did it because they could. The company held a virtual monopoly on the American telephone service, which meant it could take a longer view than corporations typically took, thinking in decades rather than quarters.

And AT&T's executives understood something that would later be

---

19 Guinness World Records. (n.d.). Most Nobel science prize winners from one laboratory. Guinness World Records. Retrieved March 11, 2026 at https://www.guinnessworldrecords. com/world-records/most-nobel-science-prize-winners-from-one-laboratory-.

forgotten – fundamental research, even research with no obvious commercial application, could yield transformative competitive advantage if you were patient enough to wait for it and wise enough to recognise it when it arrived. Their monopoly status led them to consider such endeavours as public good, actively helping the likes of Sony, Texas Instruments and IBM to commercialise their work on inventions like the transistor – those that would later become competitors as the government privatised their state-owned infrastructure.

But Bell Labs wasn't an outlier. Xerox built the Palo Alto Research Center (PARC), where scientists invented the graphical user interface, the computer mouse, Ethernet networking and the laser printer. IBM maintained research facilities where fundamental contributions to computing and materials science emerged alongside more immediately practical work. DuPont's central laboratory transformed the science of chemistry into an industrial force.

And corporate labs led to Nobel Prize-winning work. Bell Labs was just a dozen years old when Clinton Davisson, one of its scientists, was awarded the lab's first Nobel Prize in Physics along with George Thomson for confirming experimentally Louis de Broglie's wave-particle theory of matter. In addition to Bell Labs's 11 Nobel Prizes, five Nobels have also gone to researchers at IBM's labs, established in 1945 to pursue what IBM itself calls 'pure science'.[20]

Compare this to today's tech leaders. As of this writing, Google is nearly 30 years old and Apple is almost 50, but none of these tech-focused giants have generated technology worthy of similar recognition. We live in a different age.

---

20   Mills, M. (2017). 'Making Technological Miracles.' *The New Atlantis*. Retrieved March 10, 2026 at https://www.thenewatlantis.com/publications/making-technological-miracles.

These corporate labs weren't peripheral operations or corporate vanity projects. Support for pure science or basic research was common at many storied corporate labs, some we've already mentioned such as Xerox, and many others as well, including Kodak, DuPont and even Exxon. These were serious strategic investments in the uncertain work of discovery, staffed by researchers who could speak both the language of fundamental science and the language of practical application. They were the bridge that Bush's contract assumed would always exist, the institutions that caught promising research emerging from universities and figured out how to carry it toward the world.

This is what made the contract work. Not the elegant theory but the institutional infrastructure that translated theory into practice. The transitions Bush assumed would happen organically actually happened because someone made them happen.

The flaw in this logic is that Bush had assumed that these institutions would always be there, operating as they always did, and that discovery would therefore naturally be caught and carried forward. For decades that assumption held. But by the time I found myself sitting across the table from the managing partner in that boardroom who did *not* want researchers upskilled, those institutions were gone.

A system in decline

And so, by the early 2000s, parts of the bridge that had once systematically carried discovery into the world had already been dismantled. What had initially seemed like an isolated change in an otherwise stable institutional arrangement was in fact starting to be understood as part of a larger technological cycle.

Innovation systems – such as the one Bush helped design – follow predictable lifecycles. New systems, technologies and entities tend to advance slowly at first, then rapidly as they take hold, before progress tapers off as the system matures. This is the S-curve that economists have traced across every major technological revolution, where it moves from installation through acceleration into maturity and eventually exhaustion.[21]

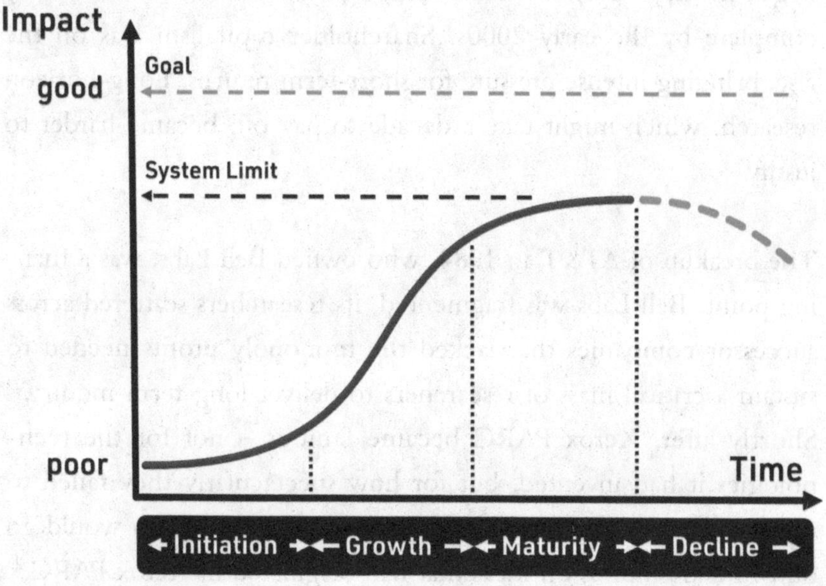

Figure: The four stages of a system's lifecycle as represented by an S-curve.

The research system that Bush designed has traced this arc precisely. Post-war expansion gave way to decades of cascading discovery and rapid growth, and then, so gradually that almost no one noticed, the plateau began. The plateau came not because researchers became less talented or because the questions became less interesting, but because

___

21 Perez, C. (2002). *Technological Revolutions and Financial Capital: The Dynamics of Bubbles and Golden Ages.* Edward Elgar Publishing.

the bridge that made the contract function was being dismantled from both ends.

The breaking of the bridge

Industry retreated first, taking its side of the bridge with it. This shift began in the 1980s, accelerated through the 1990s and was largely complete by the early 2000s. Shareholder capitalism was on the rise, bringing intense pressure for short-term returns. Long-horizon research, which might take a decade to pay off, became harder to justify.

The breakup of AT&T in 1984, who owned Bell Labs, was a turning point. Bell Labs was fragmented, its researchers scattered across successor companies that lacked the monopoly profits needed to sustain a critical mass of researchers to deliver long-term inquiry.[22] Shortly after, Xerox PARC became famous — not for the technologies it had invented, but for how spectacularly they failed to capture their value. Instead, Apple, Microsoft and Cisco would go on to build empires on ideas that had originated in Xerox PARC.[23] Pharmaceutical companies would also slowly start to shift their strategies from internal discovery toward acquisition, letting universities and small biotechs increasingly bear the early risk and then simply buying whatever emerged and looked promising.[24]

---

22   The Editors of Encyclopaedia Britannica. (n.d.). 'Bell Laboratories | History & Facts.' Britannica Money. Article last revised by Doug Ashburn. Retrieved January 5, 2026 at https://www.britannica.com/money/Bell-Laboratories.

23   Lazonick, W & Mazzucato, M. (2013). 'The risk-reward nexus in the innovation-inequality relationship: who takes the risks? Who gets the rewards?.' *Industrial and Corporate Change*.

24   Scannell, J et al. (2012). 'Diagnosing the decline in pharmaceutical R&D efficiency.' *Nature Reviews Drug Discovery*.

Seeing an opportunity, management consultants arrived in force shortly after, urging firms to focus on their core business and outsource everything else that either didn't make them money, or that someone else could pay for. Research that didn't promise near-term returns was increasingly seen as expendable, particularly by executives under pressure to cut costs and boost margins. The system as we knew it had started to crumble.

Beneath all these surface explanations lay a single, seductive logic. Basic research is expensive. It's uncertain. It's slow. It ties up capital for years, sometimes decades, before producing returns that are themselves unpredictable. Why should any corporation, accountable to shareholders quarter by quarter, bear those costs and risks when universities, funded by taxpayers, were already doing the early work?

The answer that emerged, gradually and then completely, was that corporations shouldn't bear those costs at all. Let the public sector absorb the risk of fundamental research. Let taxpayers fund the long, uncertain years of exploration and dead ends and incremental progress. Then, once promising discoveries emerged, once someone else had done the hard work of proving that the science actually worked, corporations could step in and acquire or license the results, capturing value that others had created while bearing none of the uncertainty that had gone into creating it.

It was the start of the privatisation of upside and the socialisation of risk. And it was an arrangement that proved increasingly attractive to executives who needed to deliver predictable returns to impatient shareholders and who discovered they could let someone else bear the uncertainty while they captured the rewards.

One by one, the laboratories that had once bridged their part of discovery and deployment scaled back their operations or shut down entirely. Industry was withdrawing from its side of the contract. But while corporations withdrew from translation, universities withdrew from their side of the bridge into something else entirely.

The trigger for universities was rankings. What began in 1983 as a simple *US News* survey of college presidents metastasised within two decades into a global ranking obsession, with ShanghaiRanking (Academic Ranking of World Universities), Times Higher Education and QS World University Rankings each measuring roughly the same underlying currency – publications and citations.

The world was also making it easier for students to consider looking at other countries' schools for their own further education. Universities began competing globally for these students, and these rankings became more than just a point of pride – they became critical for their growth. Studies show a 1% increase in applications for every one-spot rise in the rankings.[25] The difference between being ranked 50th and 30th could mean tens of millions of dollars in additional revenue from international students – revenue that had become essential as government funding tightened and institutions grew increasingly dependent on fee-paying students.

Rankings weren't vanity metrics any longer. They were survival.

And so universities optimised for what rankings measured, because that is what rational institutions do when their survival depends on a

---

25 O'Shaughnessy, L. (2014). 'How important are college rankings?' *CBS News*. Retrieved February 4, 2026 at https://www.cbsnews.com/news/study-highlights-influence-of-college-rankings/.

metric. They hired researchers who published prolifically. They built incentive structures that rewarded citation counts. They constructed promotion criteria around outputs that could be counted and compared. The business theorist Eli Goldratt captured the dynamic perfectly: 'Tell me how you will measure me, and I will tell you how I will behave.' The system got exactly the behaviour it incentivised.

The problem is that these metrics didn't necessarily measure things that mattered to students, to researchers or to the wider world. Publications could be counted. Citations could be counted. Teaching quality couldn't, not easily, so it slipped down the priority list. Community engagement couldn't, so it became a box to tick rather than a mission to pursue. And translation, the work of connecting research to the world, couldn't be measured at all.

The years a researcher might spend working with industry partners, navigating regulatory pathways, adapting findings for real-world conditions – none of that fed the metrics that determined whether you got promoted or whether your university climbed the rankings. Translation became invisible and unpaid work, and when it wasn't invisible it was worse than invisible, because time spent on translation was time stolen from the work that actually got measured.

A researcher who publishes 10 papers a year looks, according to the metrics, like a better performer than a researcher who publishes three papers and spends the rest of their time helping those findings reach patients or policy or practice. The rankings can't see the difference between a paper that sits in a journal and a paper that changes how a disease gets treated. They see only that one researcher produced more countable units than another.

The researchers who succeeded in this environment weren't gaming anything. They were doing exactly what they'd been told to do, responding rationally to a system that told them, clearly and repeatedly, what success looked like. The ones who won grants, who earned promotions and who built careers that their institutions celebrated, had done precisely what was asked of them. The fact that this excellence no longer connected reliably to translation wasn't their failure.

And so, as industry withdrew from early-stage research because they thought others would bear the cost to de-risk the science first, universities withdrew in the opposite direction. They retreated into the prestige economy, optimising for metrics that measured publication rather than real-world impact. Both sides had drifted from the contract Bush had brokered. While each did so for their own rational reasons, what remained was a valley the old bridge could no longer reach across.

No one had deliberately intended to break the bridge. Shareholders wanted quarterly returns. Administrators wanted higher rankings. Researchers wanted to protect their careers. Each of these choices made sense on their own. But together, without anyone intending it, they dismantled the system that had turned discovery into public benefit for half a century. The bridge didn't collapse overnight. It just stopped being able to carry the weight being placed on it, while the gatekeepers who controlled access learned how to profit from the scarcity the broken bridge created.

Researchers started referring to this as the 'valley of death'. The phrase had floated around venture capital circles for years, describing the gap between early funding and profitability that killed so many promising startups. But in research, it took on a darker meaning. It became aligned with the idea that this was the space where promising science went to die completely, and not because the discoveries lacked value or the researchers lacked dedication, but because the infrastructure (our bridge) that had once carried work across had been dismantled from both ends, and nothing had risen to replace it.

Venture capitalists sensed an opportunity. But venture capital had never been designed to fund research translation. It was engineered to fund technology companies. For decades, its model had been shaped by the software industry, a place where startups could be launched cheaply, where most ventures failed fast and where the rare winners could generate returns so enormous that they paid for all the failures many times over. But by the early 2000s, the software industry was getting crowded. So these venture investors started looking for new frontiers. Often this meant harder problems in less crowded industries where their capital might find better odds of profit-returning success.

They found exactly what they were looking for in the gap that universities and industry had jointly created. Life sciences, materials science, energy technology, climate solutions – deep technologies – these were fields where serious fundamental research was being done in universities, funded by taxpayers, but where the translation infrastructure had simply collapsed. Discoveries with genuine commercial potential were accumulating on the wrong side of the valley. And the researchers who had made those discoveries had no understanding of what they were sitting on or what it might be worth. The gap was wide open and ready for venture capital to step in and benefit.

The imbalance of information was huge. Venture investors understood the business of translation – markets, deal structures and what comparable transactions were worth. On the other hand, while researchers understood their science, often better than anyone else on earth, few of them knew anything about everything else that was required to turn it into something usable.

This was what the managing partner had described in that board-room – venture capital had learned to exploit research with precision, and didn't want to lose that edge. Researchers arrived at the threshold of translation with extraordinary expertise in one domain (research) and almost none in the domain that would determine whether their work reached the world (industry). So this meant that researchers were stuck negotiating from ignorance against parties who under-stood exactly what was on the table. They accepted terms because they had no framework for evaluating whether or not they were good for them. And they ended up handing over assets whose value they didn't understand, giving venture capitalists the edge.

And when researchers did make the deal, even the terrible deals, they were often grateful, because something seemed better than nothing and because any path across the valley seemed preferable to watching their work disappear into journals.

For a time, the arrangement appeared stable. The research system kept producing discoveries. Various forms of capital kept carrying at least some of them across. Universities kept training researchers as if the old contract still held, and enough work found its way to transla-tion that no one questioned whether the system was sustainable.

But as Sarah found out, a plateau isn't the same as stability, and it wasn't long before research institutions across the globe were entering a crisis.

The research institution crisis

This wasn't an acute crisis like a sudden collapse. Instead, it was more akin to being boiled in a slowly heated pot. Over a period of many years, government funding flattened or declined while international student markets, which had been the revenue stream masking the underlying weakness of the system, grew volatile. Meanwhile, the cost of conducting research kept rising even as the margins kept thinning.

Universities responded the only way they knew how, by pleading for increased government support, courting philanthropists with growing desperation, replacing permanent positions with short-term contracts that shifted risk onto researchers themselves and squeezing more from less, all while the underlying economics continued to deteriorate.

And through all of this, these same researchers kept producing assets, discoveries and patents, while training researchers and watching the value flow elsewhere.

The institutions and governments that paid for discovery weren't the ones seeing the rewards when that work was turned into real-world outcomes. Instead, they were propping up an ecosystem that took value from them while giving very little back.

The researchers inside these institutions experienced the same dynamic at the individual level. They created value they couldn't capture. They spent years, sometimes decades, developing expertise that existed nowhere else on earth, or generating discoveries that could transform patient lives or reshape industries. And then, when they arrived at the threshold of translation, they found themselves knowing everything about their science and nothing about what it was worth.

This information asymmetry wasn't incidental. It was structural. It was produced by an ecosystem that still assumed the old contract held and by incentive structures that rewarded publication and punished everything that didn't feed the metrics.

Building a new bridge

I thought about what the managing partner had said as I drove home that afternoon.

He wasn't wrong about the dynamics, because researchers did treat venture capitalists like sharks in the water. They'd all heard the stories about colleagues who had signed away too much, or about discoveries that had made fortunes for everyone except the people who had made the discoveries. The suspicion was rational, even if it was also paralysing.

And he wasn't wrong that the market was too thin, because when the only path across the valley now looked like a trap, most researchers didn't take any path at all, staying on their side, applying for grants and publishing their papers, watching their work slowly become irrelevant at the bottom of a pile of academic journals.

However, it was the last thing he'd said that stayed with me. Bring us competition.

What he'd meant by competition was more venture capital. He wanted more players in the same game competing for the same deals, because he believed this would get researchers moving on their discoveries, which meant more opportunities even if it compressed his margins in the short run. He was looking for a bigger pie baked from the same recipe.

I'd already seen hints that something different was stirring. I-Corps, the NSF programme in the US, and Innovation-to-Commercialisation of University Research in the UK were teaching researchers to test demand before they sunk everything into their research or product. They were showing them how to talk to customers before they wrote grants. These programmes weren't venture capital but something else entirely. A different approach to translation that was producing researchers who thought differently about where their work might go.

His conversation added another piece, because the asymmetry he'd described so candidly wasn't a bug in the system. It was a feature – one that worked precisely because researchers had few alternatives. But what if other alternatives existed, and what if someone was already building them? So I started looking for those alternatives.

I started by looking for anyone else who was having similar conversations with researchers. That search led me to the disease foundations – organisations that had spent decades funding research and watching it disappear into the Valley of Death. They had seen the same gap from a different angle, and some of them had decided to stop waiting for someone else to close it.

The managing partner had asked for competition – more players in the same game or a bigger pie baked from the same recipe. What the disease foundations had decided on was something else entirely. And what they were building was real competition; not simply more players in the same game, but an entirely different game.

They weren't just rebuilding the bridge, although they were stepping in to prop it up where they could. They'd started building a different bridge altogether, one that operated on different logic from the ground up. The old system worked because researchers had no leverage, because the information asymmetry was vast and because the only way across the valley was through gates financiers and government granting bodies controlled. Opacity that served the gatekeepers. Scarcity was a feature, not a bug.

The foundations, on the other hand, had no interest in gatekeeping. Their success wasn't measured in papers published, deals closed or returns generated, but in patients helped and diseases treated. They saw success when they could close the gap between discovery and deployment, between bench and bedside. They needed research to cross because that was literally why they existed. Which meant they had every reason to make the crossing possible, to make requirements clear and legible and to help researchers build the capability that would carry them across, rather than exploiting the fact that most researchers had never been shown.

This wasn't a better way to extract value from researchers. This was an entirely new bridge that gave these progressive foundations a better way to align the world of the researcher with theirs to help them both succeed. And they had proof it worked.

Chapter 3

THE WORLD THAT STOPPED WAITING

In 2017, the CEO of one of Australia's leading medical research foundations made a confession that stopped me in my tracks.

'We've backed research with tens of millions of dollars over the last 30 years,' he said. 'We have nothing to show for it.'

He wasn't talking about failed science. The publications were there, appearing in respected journals, accumulating citations and advancing knowledge in ways that other researchers recognised and built upon. The researchers his foundation had backed had built careers and won awards and earned the recognition of their peers. By every metric, the traditional system used to measure success, the investments had succeeded brilliantly.

But the patients his foundation existed to serve remained no closer to treatment than they had been when the funding started. The science was landing in journals, but it wasn't landing in clinics.

So, his foundation started experimenting with a different approach, and I'd been brought in to help build it. This was a more hands-on model that would scout for promising research and coach researchers to test for real demand and make meaningful and measurable progress toward translation. Over the following two years, we ran programmes together, and it started working. Researchers who came through the programme were having different conversations and were beginning to find paths the traditional system had never shown them.

This made me wonder whether we'd stumbled onto something others had already discovered. The foundation's budget was modest compared to the giant disease foundations in the US. If this approach worked at a small scale with limited resources, what might be possible with real capital behind it?

The answer, it turned out, was that a great deal was possible. And large foundations *were* already trying this approach. They had proof that made our early experiments look like pilot studies.

The problem foundations couldn't ignore

Disease foundations exist because patients and families, desperate for treatments that don't yet exist, pool their resources to accelerate research. They give in good faith to the system that Vannevar Bush designed – the research pipeline that promises discovery will eventually yield treatment, that funding excellent science means funding hope. So the foundations, backed by real people and families desperate for hope, hold fundraising galas and charity runs and memorial donations, year after year, trusting that their money is bringing cures closer.

And for decades, that trust was rewarded with publications and press releases and the assurance that progress was being made and that cures were coming closer. The foundations kept funding. The researchers kept publishing. But the patients just kept waiting.

This wasn't a failure of generosity or dedication. It wasn't a lack of research, or even good research. The infrastructure that was supposed to carry the discoveries out of the lab to the patients had collapsed. Foundations kept supplying more of the early funding that Bush had promised would naturally lead to application, and kept watching their investments evaporate in the space between discovery and deployment.

By the early 2000s, the most clear-eyed foundation leaders had concluded that waiting for someone else to fix this was no longer an option. The traditional system wasn't going to repair itself, not on any timeline that mattered to the patients who were dying while everyone waited.

So some of them started building around it.

In the spring of 2019, I was invited to lead a series of roundtables that FasterCures was convening with senior foundation leaders at the Milken Institute Global Conference. The setting was a conference at the Beverly Hilton, at an annual gathering where finance and philanthropy and policy intersect in ways that can feel surreal to anyone outside this world. The Golden Globes are held in the same hotel, which gives you a sense of the venue, though the people in this particular room weren't there to celebrate anything. They were there because they shared a common problem none of them had been able to solve alone.

The organisations represented around that table read like a directory of the most influential disease foundations in the world – the Leukemia & Lymphoma Society (LLS), the Prostate Cancer Foundation (PCF), the Juvenile Diabetes Research Foundation (JDRF), the Melanoma Research Alliance(MRA) and the Alzheimer's Drug Discovery

Foundation, along with representatives from the Gates Medical Research Institute, the Helmsley Charitable Trust and a dozen others whose names patients knew and whose funding researchers coveted.

Between them, these organisations controlled hundreds of millions of dollars in annual research funding, and they had spent two decades building infrastructure that the traditional research system had failed to provide. They had FDA approvals to prove their approach worked, drugs on the market that were extending and saving lives and billions of dollars deployed into research that had actually reached patients. By any measure that mattered to them, which is to say measures involving actual human beings getting actual treatments, they had succeeded where the traditional system had stalled.

And yet they had hit a wall.

'We've got the capital,' one CEO observed, with the kind of frustration that comes from having solved the hard problem only to discover an even harder one waiting behind it. 'We've got the structures. What we don't have is enough researchers who know how to work this way.'

The constraint they faced wasn't money. They had plenty of that and were actively looking for places to put it. It wasn't infrastructure either, because they had spent years building exactly the infrastructure the traditional system lacked. And it certainly wasn't an appetite for risk, because these were organisations that had bet hundreds of millions on unproven approaches when the pharmaceutical industry wouldn't touch them.

The constraint was the researchers themselves. Having been trained under the old Bush contract, there weren't many researchers who were ready to join them, who understood that translation was no longer someone else's job but something they could actively help with, and who had developed the necessary skills and mindset to carry their discoveries toward the world rather than simply publishing them and hoping for the best. In the language of venture capital, this was called a 'deal flow' problem. What this really meant was that they'd built the bridge from their end but almost no one was equipped or prepared to walk across.

These foundations had actually figured out how to get research to patients. They had proof that it would work. And now they were just desperate to find more researchers ready to work the way their model required.

The flaw in the traditional research pipeline

To understand why their approach worked when the traditional system didn't, it helps to know about a book that almost no one has read.

In 1997, a political scientist named Donald Stokes published *Pasteur's Quadrant*. He died shortly before it was released, and the book itself has remained an academic curiosity, cited occasionally by policy scholars and largely ignored by everyone else.[26]

But Stokes had identified something that the foundation leaders in that LA conference room would spend two decades rediscovering through trial and error, at a cost of millions of dollars and countless hours of frustration.

---

26 Stokes, D. (1997). *Pasteur's Quadrant: Basic Science and Technological Innovation*. Brookings Institution Press.

Bush's 1945 model, the model that still shaped how most researchers thought about their work, divided research into two camps. Camp one was basic research, which was curiosity-driven, focused on fundamental understanding and almost wholly indifferent to practical application. Camp two was applied research, which was practical, focused on use and almost wholly indifferent to fundamental theory.

The assumption embedded in this two-camp division was that these were sequential stages in a pipeline. First you discovered, driven by pure curiosity to find answers to questions simply because they were interesting, not because they'd necessarily lead to anything useful. The second part of this pipeline came up later and elsewhere, when someone else applied what you'd found. In this assumption, the researcher's job was discovery, and discovery only. Translation was someone else's problem.

Stokes noticed that the scientists who had actually changed the world didn't fit this approach at all.

Louis Pasteur wasn't doing abstract science when he investigated why wine and beer spoiled. He was trying to solve a practical problem that French vintners and brewers desperately needed solved, a problem with obvious commercial and economic implications. And yet his answer wasn't just a narrow technical fix. He discovered germ theory – the overall understanding that microscopic organisms caused fermentation and disease, an insight as fundamental as anything in the history of biology.[27]

---

27 Pasteur Brewing. (n.d.). Pasteur's study of fermentation. Retrieved January 3, 2026 at https://www.pasteurbrewing.com/fermentation/.

Pasteur wasn't worried about a pipeline. He wasn't choosing between depth and relevance or between fundamental understanding and practical application. He was holding both at once, letting a real-world problem drive inquiry that revealed fundamental truths about how the natural world actually worked.

Pasteur wasn't the only one changing the world with a practical approach to research and translation. Maria Skłodowska-Curie found radium and polonium while investigating radioactivity, Jonas Salk developed the polio vaccine while responding to a public health crisis, Alexander Fleming discovered penicillin during his work on staphylococcal bacteria. When Stokes examined the history of transformative science, he saw the same pattern appearing again and again.

The researchers who had mattered most, whose work had most profoundly changed human life, were not the ones who pursued pure curiosity in isolation. And they weren't the ones who focused narrowly on application without caring about underlying mechanisms. They were the ones who refused to choose, who instead embraced fundamental inquiry and practical purpose together from the start.

Stokes saw this. And he created a simple diagram, known as Pasteur's Quadrant, to capture this insight.

Pasteur's Quadrant

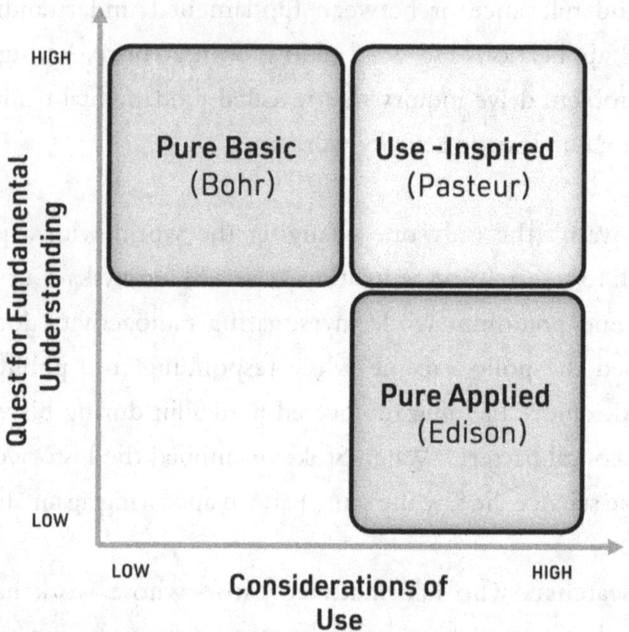

Figure: Pasteur's Quadrant

In this diagram, the upper left quadrant represents pure basic research. This is the kind of work Niels Bohr did when he investigated atomic structure out of a sheer curiosity about how the universe worked and with no thought of application whatsoever. The lower right represents pure applied research, the kind of work Thomas Edison did when he tinkered his way toward practical inventions without much concern for underlying scientific principles. But it's the upper right quadrant, the space where both fundamental understanding and consideration of use were high, where the most transformative science has happened. Stokes called it 'use-inspired basic research' and Pasteur was its exemplar.

The disease foundations that I was speaking to hadn't read Stokes's book. Most of them had never heard of him. But they had arrived at the same conclusion in a very expensive manner – by burning through tens of millions of dollars and watching carefully to see where the fires caught and where they didn't.

What they found was that the research that actually reached patients wasn't pure research that waited for someone else to handle translation. It was research that embraced from the beginning what translation would require. This was research that kept the patient in view, even while pursuing fundamental questions. It held depth and direction together, rather than treating them as sequential stages handled by different people in different places.

The most successful foundations had learned to look for this dual focus in the research they'd fund, and in the researchers they'd backed. They'd learned, just as Stokes posited all those years before, what really worked.

The foundations that figured it out

The Cystic Fibrosis Foundation (CF Foundation) had been among the first to act on this understanding, even if they wouldn't have described it in Stokes's language.

In 1989, researchers identified the gene responsible for cystic fibrosis (CF), a discovery that made headlines around the world. This result seemed to promise that effective treatments were finally within reach, because the genetic cause was now known, gene therapy was advancing rapidly and the path forward appeared clear.

A decade later, the discovery remained just that... a discovery. Patients were still dying young, still struggling to breathe, still spending hours each day on treatments that managed symptoms without addressing the underlying disease. The science had advanced enormously. But the treatments hadn't.

Robert Beall, the foundation's CEO, understood the problem. The science was there, waiting to be developed into drugs. But the market wasn't attractive enough for pharmaceutical companies to bother putting in the resources. 70,000 patients worldwide, the total CF population, wasn't a market worth chasing for companies that measured opportunity in billions of dollars. These numbers were just a rounding error on a spreadsheet to big pharma – a population too small to justify the hundreds of millions required to develop a new drug.

Beall cold-called pharmaceutical company after pharmaceutical company, asking them to develop treatments for CF. They all said no, politely but firmly, saying the science might be promising but the economics didn't work.

So the foundation made a decision that violated every norm of traditional philanthropy. In 2000, they invested $40 million USD, nearly half their annual research budget, into a small biotechnology company called Aurora Biosciences, which would later become Vertex Pharmaceuticals. The investment would fund the development of drugs targeting the underlying genetic cause of CF – not just the symptoms, but the root mechanism that made patients sick.[28]

---

28 Cystic Fibrosis Foundation. (n.d.). 'Our venture philanthropy model.' Retrieved December 11, 2026 at https://www.cff.org/about-us/our-venture-philanthropy-model.

And crucially, the foundation structured the deal so they would share in the financial returns if the drugs succeeded. This wasn't a grant that would disappear into the system and be forgotten. It was a bet, with real stakes and real potential upside, structured so that success would generate resources for more research rather than just another line on an annual report.[29]

The logic was simple, even if it seemed radical at the time. If no one else would carry the science forward, they would carry it themselves, maintaining involvement all the way through development. They would see this through to the end, rather than handing off and hoping.

So they asked themselves the first most important question – will this approach work at all? Can a foundation actually fund a drug into existence? The answer turned out to be a resounding 'yes'. In 2012, the FDA approved Kalydeco, the first drug to treat the underlying cause of CF rather than just its symptoms. Then came Orkambi, and then Symdeko and then Trikafta, which by 2019 could effectively treat up to 90% of CF patients, and effectively transform a death sentence into a manageable condition.[30]

But the real answer to that first question came when a woman named Emily Schaller, diagnosed with CF at 18 months old and told she had almost no chance of surviving to middle age, started training for marathons in her 30s and opened a retirement fund she had never imagined needing.

---

29 Cystic Fibrosis Foundation. Our venture philanthropy model; López, J & Suojanen, C. (2019). 'Harnessing venture philanthropy to accelerate medical progress.' *Nature Reviews Drug Discovery*.

30 Boyle, M. (2020). 'A message from our CEO.' Cystic Fibrosis Foundation. Retrieved December 11, 2026 at https://www.cff.org/node/871.

The foundation's original $40 million USD investment eventually returned over $3 billion USD, money that flowed back into research funding for the next generation of treatments.

The financial return was extraordinary. But the real return – the real win – was for Emily and thousands of patients like her. These were patients who could suddenly plan for futures that had seemed impossible, children who might grow up to have children of their own, families who could stop bracing for loss and start planning for life.

What the CF Foundation proved couldn't be unproven. They had demonstrated, with billions of dollars in returns and the transformed lives of patients, that the valley could indeed be crossed another way.

From one bet to a movement

Once one foundation had shown that they could pursue both research and translation, others followed the approach.

The Leukemia and Lymphoma Society (now rebranded as Blood Cancer United) launched its Therapy Acceleration Program in 2007. Since its establishment, it has invested more than $130 million USD across 70 projects,[31] including early investments in CAR T-cell therapy, the revolutionary approach that engineers a patient's own immune cells to attack cancer and that would go on to produce remissions in patients who had exhausted every other option.[32]

---

31 The Leukemia & Lymphoma Society. 'Therapy Acceleration Program.' Retrieved December 11, 2026 at https://www.lls.org/research/therapy-acceleration-program.

32 Greenberger, L. (2020). 'The evolution of CAR-T therapy.' HealthWell Foundation. Retrieved December 11, 2026 at https://www.healthwellfoundation.org/realworldhealthcare/the-evolution-of-car-t-therapy/.

Taking the same approach, the PCF raised over $800 million USD and funded research leading to 16 FDA-approved treatments, maintaining strategic involvement throughout development rather than simply writing cheques and waiting.[33] JDRF (now rebranded as Breakthrough T1D) transformed itself from a traditional grant-maker into 'a strategically-focused, priority-driven international research organization' that funds research at all stages, influencing all stages of the R&D pipeline.'[34] In 2017 Breakthrough T1D launched the T1D Fund, a hybrid structure where investment returns are recycled back into the research mission.[35]

The pattern repeated across disease after disease, from Parkinson's to multiple myeloma to Alzheimer's to melanoma, with each foundation arriving at the same conclusion through the same painful process. The traditional grant process was inhibiting real treatments from reaching real patients. They needed to build the infrastructure themselves.

Following that meeting in Beverly Hills, FasterCures would go on to build a network of more than 180 disease foundations committed to this transformation.[36] What had started as a radical bet by a single foundation had become an ecosystem, representing billions of dollars

33 Prostate Cancer Foundation. (2019). 'The Prostate Cancer Foundation Announces A Collaborative Grants Program In Prostate Cancer Research.' *PR Newswire*. Retrieved December 11, 2026 at https://www.prnewswire.com/news-releases/the-prostate-cancer-foundation-announces-a-collaborative-grants-program-in-prostate-cancer-research-300925874.html.

34 Insel, R, Deecher, D and Brewer, J. (2011). 'Juvenile Diabetes Research Foundation: Mission, Strategy, and Priorities.' *Diabetes*.

35 Kowalski, A. (2019). 'Biography of Aaron J. Kowalski, Ph.D., President and CEO, JDRF.' [Hearing witness biography]. U.S. House Committee on Energy and Commerce, Subcommittee on Health. Retrieved March 6, 2026 at https://www.congress.gov/116/meeting/house/109583/witnesses/HHRG-116-IF14-Bio-KowalskiA-20190604.pdf.

36 FasterCures, Milken Institute. 'The Research Acceleration and Innovation Network (TRAIN).' Retrieved March 6, 2026 at https://fastercures.org/programs/train/.

in deployed capital, dozens of FDA approvals and countless patients whose lives had been extended or saved by research that the traditional system would never have carried to completion on its own.

What emerged over those two decades of experimentation by these progressive foundations wasn't a collection of isolated experiments but something closer to a movement. And it was a movement that was spreading through the disease foundation world the way successful strategies always spread, by imitation and adaptation and the gradual accumulation of proof that made what at first seemed unconventional obvious in retrospect.

The bridge is built – but researchers need to cross it

When I spoke with these foundation leaders, their language revealed how far they'd moved from traditional grant-making. They talked about being 'neutral players bridging across the ecosystem', by which they meant connecting researchers with companies and regulators and clinicians and everyone else whose involvement was necessary to carry science to patients. They talked about bringing 'a sense of urgency and problem-solving across the continuum from discovery to access', by which they meant staying involved through every stage rather than funding research and walking away. They talked about being 'nimble, flexible and willing to take risks', by which they meant moving faster than government funders, betting on approaches that traditional grant committees would reject as too uncertain and structuring deals in whatever way made the science most likely to reach the people who needed it.

When I asked what they aspired to become, the answers were even more striking. 'Catalysts for disruptive, systemic change,' one CEO said, and the others nodded. Another added, 'We see ourselves as risk capital.'

They didn't think of themselves as traditional philanthropists writing cheques to support good work and hoping for the best. Instead, they saw themselves as builders of infrastructure, architects of a system that the traditional research enterprise had failed to create.

They had the capital. They had the proof. They had billions ready to deploy and a track record of FDA approvals and transformed patient lives to demonstrate that their approach worked.

And now they had a different problem. They couldn't find enough researchers to fund.

The traditional system stubbornly keeps producing scientists trained for a world that is a shadow of its former self. These are researchers who publish in high-impact journals and win competitive grants and build careers their institutions recognise and reward and who continue to pursue fundamental understanding without asking who might truly need what they discover… and who continue to struggle to make any real impact. These researchers aren't failing by any measure the system uses but succeeding brilliantly at exactly what they've been trained to do.

The problem is that their training was designed for an era when the Bush model continued to hold sway, the likes of Bell Labs still existed and industry maintained vast research operations ready to catch promising science and carry it forward. And while that era ended decades ago, the training hasn't caught up, so researchers remain stuck on one side of a broken bridge, unable to see the new bridge funders have built.

Researchers, for the most part, don't really see this new bridge – or if they do, don't understand what needs to be done to cross it. They're still trying to succeed inside a system that stopped working before many of them were even born, still unaware that an alternative has been built by people who are actively searching for them.

The old bridge is familiar and well trodden, but it's also overcapacity, overcrowded with researchers who are still struggling to push across it despite the infrastructure on both sides largely breaking down. The new bridge is younger, less visible and currently being built by foundations, philanthropists and a new generation of funders who refuse to wait. The new system operates on different logic. It's smaller, but it's growing. And it has something the old bridge cannot offer – requirements you can see, milestones you can work toward and funders on the far side who succeed only when you do.

For researchers who can see both bridges, the choice becomes clear, because only one system is on the rise. Only one bridge will help them reach their goal.

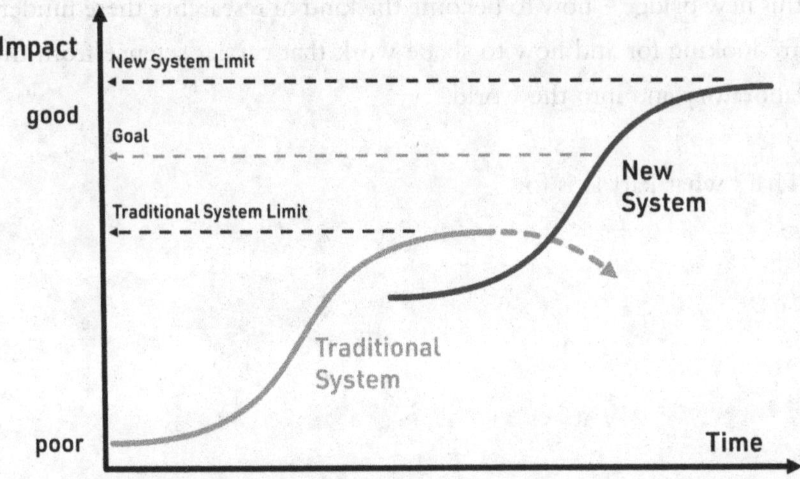

Figure: S-curve of the new system rising toward higher impact

Researchers who recognise what's happening and act on it, find what they've been looking for. They find a new ecosystem of partners who aren't looking to exploit information asymmetry. Partners whose entire purpose is carrying research to impact. And partners who succeed only when the researchers they back also succeed. The incentives are aligned in ways the old system never managed. And unlike the old system, where quarterly reports increasingly drove quick wins, those funding the research are willing to wait long enough to support real progress.

This is the researchers' real *breakthrough*. It isn't a scientific idea. It's a shift in understanding; the recognition that two bridges exist, that the one they were trained for is failing, and that the other one operates on different logic entirely, and it's waiting for one thing – researchers who know it exists and are ready to step onto it.

What remains is the practical question of how to navigate and cross this new bridge – how to become the kind of researcher these funders are looking for and how to shape work that carries science from the laboratory and into the world.

That's what Part II is for.

PART II

NAVIGATE

The new bridge exists, and the funders who are building it are searching for researchers to cross it. The question now is how you, as a researcher, become equipped for what the crossing requires.

The researchers you'll meet in this section each made this crossing starting from a different point, each met a unique set of challenges and each journeyed along a different path to cross that bridge. Some had decades of experience behind them. Others were just starting out. None of them started with an advantage that you don't have. What they had was a willingness to test their assumptions against reality, rather than waiting and hoping for the world to change, and the nerve to take the first steps without knowing where those steps would lead.

Your starting point – the specifics of your work, your context and your constraints – will be different from theirs, but there's a common pattern beneath these stories, and once you can see it, you can use it.

This turns out to be less about acquiring new skills and more about learning to work with forces impacting you and your work that have always been there, but are only visible when you know how to look.

We'll start with a researcher who had everything the old system said he needed, except a way to move forward.

The Breakthrough Model

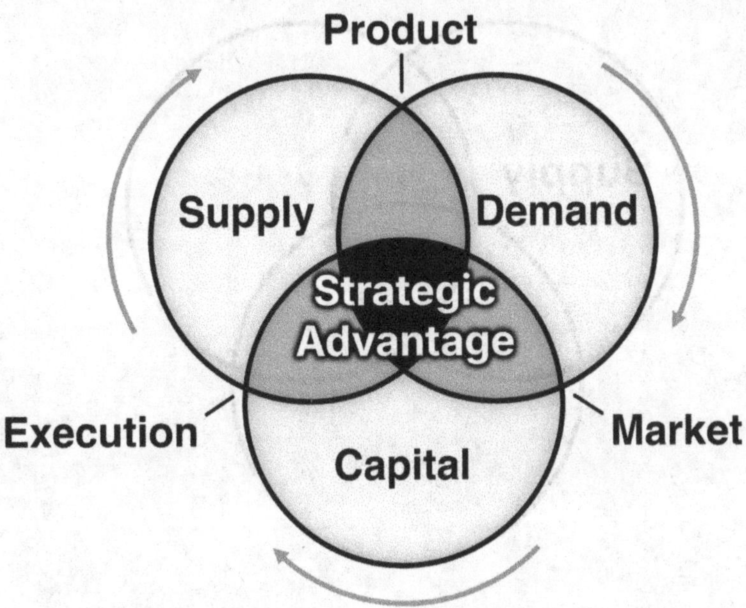

This is the map of the territory ahead. Three forces are always operating on your research whether you can see them or not: Supply, Demand and Capital. But reading those forces isn't enough on its own, because the landscape doesn't tell you what to do in response. That's where Product, Market and Execution come in, the three levers you can pull to move your work to where it matters. Part II takes you into that territory, alongside researchers who uncovered each piece, one assumption at a time.

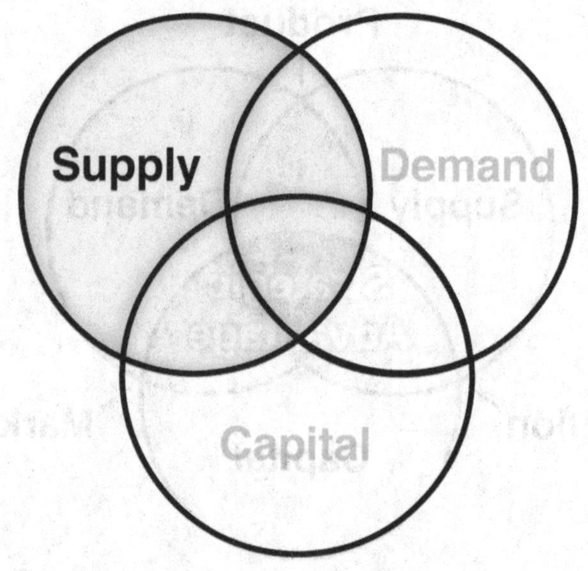

Supply is what the traditional system rewards you for acquiring and building. This is your body of work. Your knowledge, publications, expertise and all the assets that accumulate through years of rigorous work. And the system reinforces itself beautifully. Every paper accepted, every grant won, confirms that you're doing exactly what you should be doing. So you keep building, because why wouldn't you?

Chapter 4

SUPPLY FEELS LIKE PROGRESS

'They don't get it.'

Marcus said it early in our first conversation, and he would return to it several times before we were done. It wasn't a complaint – he was genuinely bewildered – exhibiting the confusion of someone who could see something clearly and couldn't understand why the world wasn't doing anything about it. He wasn't talking about his peers, who understood the science perfectly well, or the journals, which had published hundreds of his papers over two decades. He was talking about the people with money. Marcus had spent 20 years researching diabetic complications. He'd published hundreds of papers, was ranked among the top experts globally in his field and he believed with absolute conviction that the inflammatory pathway he'd mapped could transform how we treat diabetes.

He could trace the mechanism precisely, could explain why it mattered for the millions of patients living with the disease and could point to the data that supported every claim. The evidence was incontrovertible. But every time he'd tried to secure the Capital to develop it further, the answer had been some version of 'not yet'. The science was too early. The indication was too broad. The pathway was too novel. Each rejection had a different rationale, but they all pointed to the same conclusion: the money wasn't coming.

Successful but stuck

Marcus had come to research through clinical practice, through seeing the devastation of what diabetes could do to the human body over time. The blindness. The kidney disease. The amputations. The threat of constant dialysis. He had entered the laboratory precisely because he wanted to change those outcomes. He wanted to build the scientific foundations for treatments that might spare future patients what he'd watched former patients endure. And this motivation never wavered.

So he did everything the traditional system asked of him. He built the publication record, established the laboratory and earned the recognition of his peers. And over the decades, the publications multiplied and the recognition grew. By every metric the system used to measure success, he had succeeded. But the gap between what he knew and what patients received remained as wide as it had been when he started. Not because people doubted his science, but because nobody in the traditional system could show him how to find funding for what had to happen next.

His statement that 'they don't get it' wasn't rooted in arrogance. It was stemming from the frustration of someone who had spent a career building something valuable, only to be told repeatedly that the Capital to develop it wasn't available and so couldn't find the path that would lead it out into the world.

Assumptions that feel like facts

What Marcus was missing wasn't more publications or more validation, and it wasn't more Capital either, though that's certainly what he believed. What he was missing was a way of seeing that the old system had never taught him. He was missing a framework for understanding why some science moves toward translation while other science, equally valid, equally important, sits in journals accumulating citations that never convert to anything beyond more citations.

While Marcus was trying to force his work through a system that had no real path for it, there were people elsewhere who were actively searching for research like his – and researchers like him.

A European investment firm that specialised in deep technology research had recently arrived in Australia. I was familiar with the team and I knew they were serious. They had Capital ready to deploy and had already committed hundreds of millions of dollars to medical research. They also had expertise in therapeutic development and access to infrastructure designed to help carry promising science across that 'Valley of Death' that swallows so much academic work. Most importantly, they needed researchers exactly like Marcus.

The problem was, the investment firm didn't know Marcus existed, and Marcus didn't know they existed either, because the old system that had trained him and promoted him and validated him at every step had never once mentioned that people and entities like this were relevant to his work.

Here was another way across that valley entirely, with people and money on the other side of it who were actively willing researchers to use the crossing. But the old system had never once pointed Marcus toward it. No signposts, no map and no one saying, 'Hey check out this other path.'

So I did. But when I raised the possibility of approaching venture capital, Marcus's response was immediate.

'No way. This isn't ready for VC.'

I persisted, and asked about the university's commercialisation team, whether they might be ok with an introduction.

'That's the other thing,' he said. 'They'd never let me go in that direction. Not at this stage.'

I asked what made him so certain, and he explained that he'd seen how these things worked. The technology transfer office controlled all commercial relationships and would never approve a conversation with outside investors for science this early. Even if they did approve it, the investors would want sure things and his work wasn't anywhere close to what serious people would take seriously.

These weren't, and aren't, unusual beliefs. They're the assumptions that stop most researchers before they even start. And it's not because Marcus, or researchers generally, are hostile to translation. It's because the old system has consistently reinforced the idea that certain paths are closed, certain conversations are premature and certain people are gatekeepers who must be appeased before any movement is possible. And sometimes, just sometimes, you can still hear the old guard from the traditional system saying that such Capital from industry or investors is 'dirty money', which doesn't help anyone wanting to explore the new system with safety and support.

For Marcus, and countless other researchers in the same position, these beliefs felt like facts because they'd never been tested. Everything these researchers knew made them believe that the university's commercialisation team would block any approach to outside Capital. That venture investors wanted certainties, not mechanisms still being validated. That if the science wasn't packaged correctly or wasn't mature enough, then it also wasn't at a stage where anyone with resources would look at it or back it.

Beneath those surface beliefs, a deeper fear lurks — one that most researchers carry but rarely name — the fear of being told your work isn't worth it. The fear of taking the science you've devoted your career to, and having someone with money look at it and say no. This can feel like a risk, and it's often easier to stay in the familiar system, where validation comes through publications and grants and conference invitations, than to risk the kind of rejection that, without the right support, can kill the motivation to keep pushing forward.

These beliefs aren't irrational. In fact, they're a rational response to a world that has never shown you anything different, that reinforces the same beliefs through training and incentive structures and daily experience until any other options become invisible, and you start treating it as simply the way things are.

Testing the assumptions

I was determined to challenge these assumptions with Marcus. So I said, 'Have you asked them? The commercialisation team. Have you actually had that conversation? And do we know the VCs aren't interested, or is that an assumption too?'

A long pause.

'No,' Marcus said. 'I haven't asked.'

'Well then. Let's test it. What have we got to lose?'

What Marcus needed wasn't to abandon his caution, which was reasonable enough given what he'd observed over two decades inside the system. What he needed was to give himself permission to test his assumptions against reality. He needed to find out whether the obstacles he perceived were actually there or were just projections of a system that had taught him to expect barriers without ever checking whether the barriers existed in his particular case, with his particular science, at this particular moment in time.

The only way to find out was to ask, and so he did. First, he spoke to the commercialisation team whose resistance he'd spent years assuming without ever actually testing. The conversation lasted 20 minutes. And the result was not at all what he expected.

Marcus was prepared for resistance. He was ready for a 'no'. After all, technology transfer offices have historically guarded commercial relationships closely, positioning themselves as the only route through which any conversation with industry or investors must flow.

I remember one meeting I attended with a researcher and their tech transfer officer, who had joined uninvited. When it came to introductions, the tech transfer officer pointed to his researcher and addressed the entire room saying, 'I'm just here to make sure he doesn't say anything stupid' – a destructive mindset for everyone around the table. I ended that meeting immediately.

Such experiences echo through the halls of research, leaving researchers believing that if they try to go it alone they can expect friction at best and blocked projects at worst. Marcus too had been told by colleagues to expect that such a conversation with his technology transfer office would end with polite discouragement and a suggestion to wait until the science was further along.

But the landscape had been shifting in ways Marcus hadn't noticed. Some business development leads had begun to recognise that researcher-led exploration of commercial interest could complement and accelerate their work, rather than threaten it. These people were beginning to see that early validation of Demand made their jobs easier because it helped to filter opportunities before they hit the formal pipeline. I had reason to believe Marcus's commercialisation lead was one of these people, and that turned out to be true.

What Marcus encountered during that conversation wasn't resistance at all. It was something closer to relief. The office wasn't at all opposed to him having exploratory conversations with investors – they welcomed it. They asked only that confidential IP remain confidential and that he keep them informed of anything substantive. Beyond that, they said, 'Go, talk, see what you learn.'

Marcus's fear wasn't unfounded. This same conversation, handled differently, at another institution, with a different individual, at some other moment in time, could have looked very different. But he had never tested whether that resistance existed here, with this person, for this project, right now. And it turned out it didn't. What Marcus had needed, before anything else, was permission to ask – not from the commercialisation office, but from himself.

Finding your market

Once he gave himself the permission to start talking about commercialisation, Marcus was off and running. We reached out to the investors who agreed to meet. This wasn't a guarantee that any progress would be made for the patients that he was so concerned for, but the fit was there on paper. He had a novel mechanism, years of rigorous validation and clear therapeutic potential across multiple indications. The conversation was there to be had, and now Marcus could see a path forward.

What mattered now was how Marcus approached the conversation itself. The instinct for most researchers in any type of funding conversation is to slip into presentation mode. They want to explain the science, walk through the data and build the case for why this work matters. It's the mode they've been trained in. It's also the format of every conference talk and grant application and journal article they've ever seen or produced. But presenting is a one-way transmission, and what Marcus needed wasn't to transmit, but to receive.

Marcus needed to structure a conversation that would allow him to find out what these investors saw when they looked at his science.

He needed to find out what applications occurred to them, what gaps they perceived and what would need to be in place for them to consider backing him.

So, in the meeting itself, Marcus did something that felt awkward but was absolutely essential – he listened more than he talked. He treated the conversation as an exploration rather than an evaluation. When they probed, he asked questions in return. He sought information about their point of view. And he made it clear that this was an opportunity for him to learn how his work appeared to them – to the people who viewed the potential and opportunities from a completely different vantage point.

He learned a great deal from this conversation because they asked him things he hadn't been asked in his decades-long academic career. Not about the mechanism itself, which they grasped quickly enough, but about applications he hadn't seriously considered. Where else might this pathway matter? Who else might be searching for approaches like this? Where might the Demand be greatest, the regulatory pathways clearest, the unmet need most acute?

The investors approached opportunities more broadly than Marcus ever had because their view was shaped by an understanding of markets, regulation and opportunities to scale, rather than by a single field of study.

What Marcus had never been taught to see, and what the investors saw immediately, was how his science fit into a larger picture. They weren't evaluating his mechanism in isolation. They were reading conditions he'd never learned to read: what capabilities existed to carry the science forward, who was actively searching for solutions, where Capital was flowing and why. These were the forces that determined whether research moved, and until that conversation Marcus had been blind to all three. This perspective meant that within minutes they'd identified something he'd been too close to see – a new market.

Marcus's focus had always been kidneys. This was the organ system his clinical experience had pointed him toward, and the complications that first drew him into research. But the pathway he'd mapped was expressed throughout the body. And lung tissue, it turned out, expressed it more highly than almost anywhere else. So they suggested chronic obstructive pulmonary disease (COPD). There were hundreds of millions of patients suffering from COPD worldwide, and there were already established regulatory precedents for the therapeutic approach his mechanism would require. In other words, there was another market actively searching for new solutions, and Marcus had one to offer.

It wasn't that Marcus didn't know his pathway was relevant to other organs. The data was there in his own publications. But he'd never asked who else might need what he'd spent 20 years building. And that's because he'd never framed the question in terms of Demand rather than scientific insight. So the lung application had remained invisible to him even as it sat in plain sight in his own research.

What struck Marcus most wasn't the new indication, though that would prove transformative. It was that these investors hadn't told him his work wasn't ready. They hadn't said the science was too early or the pathway too novel. They had looked at his research – the same mechanism the traditional system thought was too early to fund – and had seen something they wanted, even if it was in a different way than Marcus expected. And they were ready to fund. The Capital Marcus had been chasing hadn't been unavailable. He just hadn't been looking in the right place.

After years of presenting his work to people who acknowledged it was important but had no way to move it forward, what happened in that meeting mattered more than Marcus had expected. The investors saw a potential foundation for therapeutics that could address inflammatory disease across multiple organ systems. And unlike the grant committees and journal editors and conference audiences who could only nod, as venture capitalists they had everything they needed to act.

But more significant than their belief was what they had shown *him* – the pathway to lung disease. And it wasn't a detour from his original vision. It was a route toward it, because proving the mechanism in one indication would validate the approach for others. Kidneys, liver, neurodegeneration, everything he'd always cared about would become accessible not *despite* taking a different pathway through lungs but *because* of it. The approach the investors identified wasn't a compromise. It was a new way to achieve his mission.

What Marcus had discovered wasn't just a better market. He'd also discovered that people existed who wanted to see him and his research succeed and had what he needed to make that happen – Capital to deploy, expertise to contribute and infrastructure to provide. More importantly, he'd found that there were people who could see applications in his science that he'd been too close to appreciate. And they weren't just offering money. They were showing him a way to cross that he hadn't known existed – a place where the science he'd spent decades building could meet the world that had always seemed unreachable.

There was a way across the gap for his work after all. And he could see that crossing now that he knew how to look.

What lay ahead remained uncertain. He didn't know whether the investors would commit, whether the science would translate or whether the lung application would prove as promising as it appeared. But for the first time in his career, Marcus could see a way forward that didn't require waiting for the traditional system to notice him.

Seeing beyond Supply

For 20 years, Marcus had been building what I introduce to researchers as Supply. Supply was everything he possessed that he could contribute to the therapeutic development of his research: the mechanisms, the publications, the expertise and the scientific validation. And as the traditional system's appetite for bold work narrowed into smaller, more incremental pieces of research that felt safe, it left work like Marcus's stranded without the support that might once have carried his science forward.

The science was good. It was genuine. It was as real as any researcher's, built through decades of rigorous work. Marcus was trapped by his unconscious focus on Supply, as the traditional system taught him to believe this was the sole measure of progress. Since that system's function was academic recognition, not development, it offered no bridge to translation. Research stalls when Supply is pursued in isolation, as it cannot create movement toward a legitimate need.

What Marcus found, once he looked beyond the traditional system, was a Demand for him and his work. Not need, which is passive, and sits there waiting to be addressed, but true Demand, which shows itself when someone with resources is actively searching for a solution to a problem they've decided to commit those resources to solve.

The investors had shown him there was Demand from venture capital, without using this language, when they showed him how his science could become something else. This wasn't about the mechanism itself, but where the mechanism could be aimed to address a specific set of patients whose need was acute and whose numbers made the economics work.

Marcus could now see that there were people out there who could help translate his science into the world. There were investors hunting for novel mechanisms, development partners scanning for validated targets and disease foundations searching for science they could carry toward patients. These weren't people who were waiting to be convinced but who were already looking, already had money allocated, already had infrastructure in place, and who were ready to move the moment they found research that fit what they were searching for.

The research world tends to assume that if the science is good enough, Demand will materialise, that the quality of the work will somehow summon the people who need it. But the disease foundations in that LA conference room, who you were introduced to in Chapter 3, weren't searching for good science in the abstract. They were searching for science that addressed the conditions their patients suffered from and was delivered in forms they could develop, by researchers who understood how to work with them.

Such Demand is always specific, even when the science it's looking for is general. Marcus had spent his career being general. But now, he was gaining insights into how to make his research specific, and this was changing his perspective as a researcher.

Marcus could now see where his Supply met a specific Demand, and at that intersection he'd shaped a product he'd never imagined he would. And because the Demand was there, Capital was suddenly within reach. He'd been building his Supply for years, unable to see why the Capital wasn't flowing. But now, with both Supply and Demand insights, he could see the solution.

While Marcus had been trapped inside Supply, the next researcher had the opposite problem. He could feel Capital pressing on every decision, every day, so acutely that it had become the only thing he could see. And what it was hiding from him was the very thing Marcus had started with.

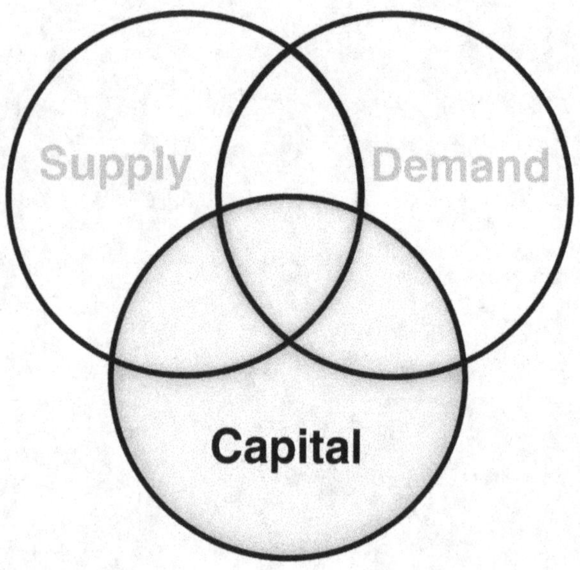

Capital is the force every researcher feels most acutely, because without it nothing else is possible. The lab needs equipment, the team needs salaries, the experiments need time, and time costs money. So you orient around it. You write grants, you chase funding cycles, you learn to shape your work around what the money will support. It feels like diligence because the system demands it. But what if it isn't?

Chapter 5

CAPITAL FEELS LIKE THE ANSWER

Kusal's computers were failing. Not in any dramatic way. No sparks or smoke or sudden black screens appeared. They were just too old and too weak for what he needed them to do.

The machines sat in the corner of someone else's laboratory, borrowed space that he used because he didn't have anywhere else to work. He would set up the computational work he needed to run the spatial analysis of tumour tissue that was his life's obsession, and the outdated machines would grind through the overnight processing jobs that made them crash more often than let them complete it. Every morning, he'd head into the lab, check the screens, restart the ones that had stalled and set the others to run the next set of processing jobs, all the time knowing that by tomorrow the same thing would happen again.

The machines weren't built for this kind of work. And they were slowing him down in ways he could feel but couldn't solve, because solving the problem required money, and money was one thing he didn't have.

This was 2020, and if you'd looked at Kusal's situation from the outside, what you'd have seen was a researcher in trouble. He'd completed his doctorate three years earlier and had won a competitive national fellowship that should have launched him into the next stage of his career. He was pioneering techniques for mapping what happens inside tumours at the level of individual cells – work that almost no one else in his part of the world was doing, and that would prove, in time, to be among the most important methodological advances in cancer biology.

But none of that mattered when the computers crashed.

Kusal was experiencing something most early and mid-career researchers would recognise instantly, even if they'd never describe it in quite these terms. He was trapped in a cycle where the science required resources, the resources required grants, the grants required time to write and administer, and the time spent on grants was time stolen from the science that justified the grants in the first place. Every week was a negotiation between doing the work and funding the work, and funding was winning, because without it nothing else was possible.

Kusal was so busy plugging holes in a sinking ship, patching one leak only to discover another, he was never able to focus on the things that would actually sail this ship to his destination, or any destination for that matter. He had even started to wonder whether the ship could be saved at all.

Kusal had begun thinking seriously about the viability of his career, because he couldn't see a path forward that didn't involve more of the same: more scrambling, more borrowed equipment, more grant applications written in hours stolen from research, more of the endless treadmill that was consuming the best years of his career without producing anything he could point to as progress. The system was doing what systems do when they're under strain. It was pushing out the people it most needed to keep.

This is what Capital looks like when it becomes the only force you can see. It shrinks your world. It reduces every question to a question of funding and every conversation to a conversation about survival. It makes you so focused on what you lack that you lose sight of what you have, and what Kusal had was extraordinary. He just couldn't see it anymore.

What Capital blindness hides

When I first started working with Kusal, I expected the conversation to be about funding. That's usually where researchers in his position want to begin, because Capital is the wound they can feel, the thing that hurts every day, the problem that seems most urgent and most tractable. *'If only I could get this grant, if only I had better equipment, if only I had more time protected from administrative burden, then the science could move.'*

But Capital, for researchers, is like taking pain medicine. It gives you relief from the symptom, but doesn't treat the underlying cause. That's because just as the real issue isn't the pain, but the wound – the real challenge for researchers isn't how to find more money, it's how to make the money find you?

This seemed to be precisely what was going on with Kusal. So, instead of starting with Capital, we stepped back from it entirely.

We spent most of our time together understanding Kusal's sources of Supply, and what emerged was something neither of us had expected.

Kusal had been describing his work the way the traditional system had taught him to: as a set of projects, each requiring funding, each competing for attention, each operating more or less independently in the constant scramble for resources. One project here, another there. Seen from the everyday reality of grant writing and funding applications, each effort appeared separate, demanding its own time, team and money.

But when we started unpacking what he actually had, what he knew, and what he was, a different picture emerged entirely. Kusal wasn't running a collection of disconnected projects. He had built, almost without realising it, a network of spatial biologists that could turn their hand to virtually anything. More than that, he'd built a series of relationships around his work. Clinicians trusted him. Technologists collaborated with him. He had relationships with instrument companies, hospital pathology departments, international researchers and patient advocacy groups. He was producing a remarkable volume of work despite conditions that would have stopped most people cold, and the reason he could produce that volume was that he'd assembled, through years of relentless, collaborative effort, a constellation of capabilities that existed nowhere else in his field.

What we discovered working together is that Kusal was sitting on what was essentially a platform. He thought he was starved of Capital. In financial terms, he was. But in every other sense – human, relational, intellectual, infrastructural – he was Capital-rich. He just didn't recognise those forms of Capital because the grant cycle had trained him to see individual projects as funding opportunities rather than a chance to leverage the underlying capability that connected them. But that's what he had.

Each conversation revealed something new he'd overlooked or undervalued or simply never thought to name. The network he'd built wasn't incidental to his research. It was his research, or rather, it was the infrastructure that made his research possible, and that infrastructure was exactly what funders, the right funders, were searching for.

Kusal wasn't alone in missing this big picture, because this is what Capital blindness does. It keeps you so focused on the resource constraint that you never step back far enough to see the asset you've built. And because you can't see the asset, you can't articulate it, which means you can't show it to the people who might want to back it, which means Capital stays scarce, which keeps you focused on the constraint. And the cycle continues and continues and continues.

Seeing your own Supply clearly is one of the hardest things a researcher can do, precisely because you're too close to it. The lens that makes you world-class in your field also makes you blind to how that field connects to everything else. Kusal knew what his science could do. What he couldn't see was how it looked from the outside, from the vantage point of someone looking for capability worth backing.

So we reframed and repackaged his Supply to be seen and understood through other people's eyes. Not his own assessment of what he'd built, which was shaped by years of scarcity and survival, but the assessment that institutes would make and that granting bodies and philanthropic organisations would make as well, if they could see the full picture rather than the fragmented version the grant cycle presented.

Rethinking how we talk about Supply

The test came at a conference in the US. Kusal had attended conferences before, of course, and he'd always done what researchers do at conferences. He'd presented his findings, attended sessions and networked in the carefully bounded way that academics network, exchanging polite interest, business cards and vague promises to collaborate. But this time was different, because this time he understood what he had to offer.

We'd spent weeks preparing for this conference. We weren't working on his slides, so much as his narrative. We worked out how he would talk about his work, introduce himself and answer the inevitable question: 'What are you working on?' We reframed the story so that it led with the problems he and his platform could solve for others, instead of simply presenting his capability and his need for further funding.

The shift sounds simple. It's not. Researchers are trained to present scientific capability and credentials, because these justify funding, and funding keeps the lights on. The new approach required Kusal to fundamentally reposition his work: instead of leading with what he did, he began with the problems he could solve together with his platform. This required him to stop saying, 'I need resources to continue,' and to start saying, 'I have built something valuable that can fix these specific problems.' He was still talking about his Supply, but he was framing it through a lens of Demand and with a new script.

As he started talking with others using this new script, it felt strange. Leading with problems he could solve required Kusal to resist every instinct his training had built, and he stumbled over the new approach more than once. Time and time again, he'd catch himself drifting back toward the old script of projects and funding gaps. It didn't feel polished or easy. It felt like he was trying to speak a new language.

But after he started using the script, even though it was awkward and unpolished, something important shifted. The response wasn't the usual polite academic interest, the kind where people nod and move on to the next session. Instead, people really listened. They sought him out. They wanted to talk, and not about his latest paper but about what his platform could do, the other problems it could be turned toward and what collaboration might look like. He was in demand throughout the conference in a way he had never experienced. He was invited to dinners he hadn't expected to attend – the kind of informal gatherings where pharmaceutical executives and biotech leaders and fellow researchers sat around a table and talked about what was coming next in the field.

When he got back home to Australia, he was buzzing. And the buzzing wasn't the fleeting high of a well-received talk or a newly published paper. It was a buzz that came from the recognition that the world had been looking for what he'd built all along. And because he'd finally let them see what he had to offer, everything changed.

This was also what Marcus discovered in his investor meeting – that everything changed when he stopped arriving to present and started arriving to listen. The conversation shifted because he realised the people across the table could see value in his work in a different way than he ever had himself.

For Kusal, the shift was different but just as profound. He didn't need to change how he spoke so much as what he understood himself to be. Whereas Marcus had always known what his science was, what Kusal discovered was who he was. Not just a researcher running projects, but the architect of an in-demand capability others would pay to access. He finally saw himself as someone able to build collaborative infrastructure from almost nothing.

Seeing your Capital focus clearly

What happened at that conference didn't solve Kusal's funding problems directly. No one wrote him a cheque over dinner. But something shifted in how he understood his own position, and that shift presented him a whole new pathway.

Kusal's story also reveals something important for any researcher looking to cross the bridge to translation, and that's the second force: Capital. Capital is the resources that enable movement from discovery toward deployment: financial Capital, intellectual capital, human capital, infrastructure. And what Kusal's experience shows is something that surprises almost every researcher who encounters it, and that is that such Capital doesn't work the way the traditional system teaches you to think it does.

The traditional system treats Capital as something you extract. You write a grant application, you submit it to a committee, the committee decides whether to give you money and, if they do, you take it and return to the laboratory. Capital, in this framing, is a transaction, but one that you're powerless in. You ask, they decide. The decision-making sits entirely on their side of the table.

Every researcher who has sat in front of a grant committee knows this dynamic intimately – the supplicant posture, the careful framing of objectives in language the committee wants to hear, the gnawing awareness that your career progress depends on people who have never met you reading a document that can't possibly convey the full depth of what you're aspiring to achieve.

But Capital, as a force shaping translation, operates on entirely different logic. Capital moves. It moves toward demonstrated value the way iron filings move toward a magnet, not because anyone commands it to, but because the people who control it – investors, foundations and philanthropists – succeed only when they find the right things to back. They aren't gatekeepers deciding who deserves support. They're searchers, scouring the landscape for capability worth investing in. Their problem is almost never that they have too many good options to choose from. Their problem is that they can't find enough of the right options that will help them succeed with their goals.

This isn't a theory about how Capital ought to work. It's a description of how the people who deploy it actually think. Cognitive scientist Lia DiBello's research on business expertise suggests that great business operators share a common tacit mental model of how business works: as an integrated flow of forces rather than a checklist of factors.[37] What Kusal was learning to do, without knowing the research behind it, was to see his own work the way those operators already saw it.

---

37 DiBello, L. A. (2019). 'Expertise in business: Evolving with a changing world.' In P. Ward, J. M. Schraagen, J. Gore & E. Roth (Eds.), *The Oxford handbook of expertise*. Oxford University Press.

This is what Kusal discovered, though it would take months for the full implication to settle. The Capital he'd been chasing, the small grants and fellowship applications and equipment requests, was transactional Capital, the kind the traditional system dispensed through committees and applications. It was real, and it mattered, but it was never going to take him where he wanted to go because it wasn't designed to. It was designed to keep the existing system running, not build something new.

The Capital that would change everything, and that would eventually find him, operated on the logic of the new bridge, though Kusal couldn't see that yet. It's attracted toward demonstrated value and toward packaged capability that could be seen and understood and backed with confidence.

But if it was always there, why hadn't Kusal found it earlier? The reason wasn't that the Capital didn't exist or that the funders weren't looking. It was that Kusal hadn't been able to articulate what he had, because Capital blindness had kept him focused on what he lacked.

From scarcity to abundance

The two years of intensive work that followed Kusal's time in the US traced an arc that would have been unimaginable when we first met, when he was sitting in the corner of someone else's laboratory.

Kusal soon moved from his original institution to a major research university that could now see his potential. He hadn't applied for a new position in the conventional sense. But because the reframed understanding of what he'd built made him visible to institutions, they now sought him out because they were looking for exactly the kind of collaborative, translational capability he represented. The move wasn't just a career step (even though he now had access to far better computers). It was evidence that Supply, when it can finally be seen clearly as a product by forces of Demand, attracts the conditions – the Capital – for its own growth.

At the new institution, the pattern accelerated. The network he'd built kept expanding. The collaborative relationships kept deepening. The funding that had seemed so scarce began arriving from directions he hadn't expected, not because the funding landscape had changed but because his positioning within it had. He was no longer a researcher scrambling for small grants. He was becoming something else entirely, though what that something else would look like hadn't fully revealed itself yet.

By late 2022, the intensive reframing had done what it needed to do. Kusal understood something that most researchers never learn: that Capital, when you stop chasing it and start demonstrating value, has a way of finding you. The scramble doesn't end completely, because research always requires resources and resources are never infinite. But your relationship with Capital changes. It shifts from extraction to attraction, from supplication to partnership and from a position of weakness to one where the people with resources need what you've built as much as you need what they can provide. What he came to truly understand was that the relationship with Capital in this new system is a two-way street, a give-get and win-win equation that must balance for the relationship to thrive.

Where this would carry him was something neither of us could have predicted. What we could see was that, beyond his work, something was starting to move for Kusal too. The first steps were hard, things felt heavy, even clumsy. But one day we realised that the man who had been checking crashed computers every morning was moving. The momentum was building and, for the first time, the world was starting to take notice of what he was building.

What's next

Two forces are now visible – Supply and Capital. But there's a third that neither Marcus nor Kusal had initially recognised – one that completes the picture and changes how you read everything else. It tells you whether what you've built connects to what anyone actually wants, and whether the Supply you've assembled and the Capital you've attracted are aimed at creating something real: Demand.

And, as it turns out, this final force is the hardest of the three to see clearly. That's because it's the one most likely to lie to you.

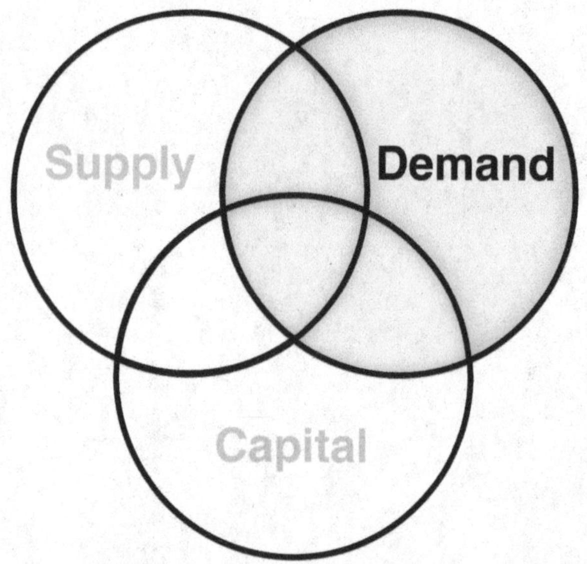

Demand seems like the simplest of the three forces. Someone either wants what you're building, or they don't. But Demand has convincing imitators, and these imitators are everywhere. Fake Demand shows up as interest, as encouragement and as people offering opinions about where it should go. It's possible to spend your career surrounded by what looks like Demand without any of it ever contributing to moving you forward. But what if there was a way to find authentic Demand, not poor imitations?

Chapter 6

DEMAND IS EASY TO MISTAKE

James's idea had taken shape during the COVID-19 lockdown in 2020, in his garage, which is perhaps a cliché for consumer software but not for the kind of deep biology James was exploring. He'd been thinking about a class of biological targets, proteins that drove disease but had resisted every conventional drug approach. The pharmaceutical industry had tried to make inroads on treatment for years, had hit wall after wall and ultimately moved on. But James believed he could see something they'd missed.

His idea felt important. It also felt impossibly large, and every time he tried to talk about it with colleagues, the conversations pulled him in a different direction.

Some thought he should explore the underlying science further, that he should run more experiments before drawing conclusions. Others wanted to leap ahead to applications and markets, impatient with the slow pace of academic discovery. His commercial team wanted to talk about competitors and revenue projections, using the language of business plans rather than lab work.

Each perspective made sense on its own, but none of them seemed to present him with a real path forward. Should the technology be a platform or should he focus on a single indication? Was it a research tool or a therapeutic programme? At the core of the problem was that he didn't know how he should be packaging his research, because he didn't actually know what, or if, Demand existed.

Without a practical foundation to stand on, James found himself pushed by whatever wind happened to be blowing on any given day. He wasn't lazy or unfocused. He simply had no anchor, no clear sense of what he actually possessed that he could move forward with, and so every new conversation became a new direction to consider. Two years passed without the idea meaningfully progressing in any direction at all.

The illusion of fake Demand

Here's what makes James's situation different from Marcus's and Kusal's, and what makes Demand the most treacherous of the three forces.

Marcus had been blind to Demand entirely. He'd spent 20 years in Supply and had never looked beyond the traditional system to see who might want what he'd built. When he finally looked, he found Demand waiting for him in places he'd never considered checking. The fix was relatively straightforward: step out of Supply, discover Demand and let the intersection of these forces reshape your initial understanding of your product.

Kusal hadn't been able to see his own Supply because Capital blindness had clouded his field of vision. The fix was to step back from the funding scramble long enough to see what he'd actually built, and then show the world what it was looking at.

James had a different problem. He wasn't blind to Demand. He was surrounded by it, or what looked like Demand, every day. Colleagues offering opinions. Advisors suggesting directions. Peers at conferences telling him what they thought the market needed. Commercial teams projecting revenue for products that didn't yet exist. There were people erecting signposts everywhere, but they were all pointing somewhere different.

The underlying issue with where James was looking for direction on Demand was that none of these people had skin in the game.

A colleague who suggests you should explore a particular application isn't showing you Demand. They're offering an opinion, and even if their opinion comes from a place of expertise, it costs nothing to give and nothing to abandon. A commercial team that projects revenue for a technology platform isn't showing you Demand either. They're modelling a hypothesis, one that will never be tested against their own resources or reputation. Even an advisor who says the field is moving in a particular direction is describing a landscape – not Demand, and because they're not committing to back your journey across it, they've got nothing to lose either.

Each of these were offering him false signals of Demand. But authentic Demand, the kind that moves research toward translation, comes from people who are willing to commit something to your project: Capital, time, access or reputation. It comes from people and organisations that have a problem they need solved and resources they're prepared to deploy toward solving it despite there being a risk involved. It comes from the willingness to act, not just the willingness to advise, and the distance between those two things is where most translation efforts quietly die.

James had been swimming in a sea of fake Demand for two years, and it had kept him in constant motion without ever moving him forward. Every conversation felt productive because someone was engaging with his work. Every piece of advice felt valuable because it came from someone who understood the field. But none of it had moved him closer to knowing what to build, because none of it came from people whose own futures depended on the answer.

This is what makes Demand the hardest force to read. Supply you can see because you built it. Capital you can feel because it's either flowing or not. But fake Demand lies. It looks like engagement when it's really just curiosity. It looks like encouragement when it's really just politeness. It looks like direction when it's really just opinion. When we're fooled by fake Demand, we don't look for real Demand. And a researcher who can't tell the difference between authentic Demand and its imitators can spend years responding to fake signals that will never convert to anything, because the people making them were never going to commit.

Finding authentic Demand

The shift came when James reconnected with Ben. They'd done their doctoral work at the same institute years earlier, but since then they'd taken different paths. While James had gone deeper into the laboratory and pure research, Ben had been working at the boundary between science and commerce, taking on roles that required him to learn how to evaluate research possibilities.

It was Ben's observations that changed the trajectory. He'd been watching the drift, watching the idea lose coherence under the weight of competing advice, and he suggested that they could explore spinning the technology out. Not because the science was ready, necessarily, but because the discipline of building something would force clarity that two more years of academic conversation never would.

That suggestion led them both into an annual programme I was running in their institute, where they were expecting to learn how to package, position and pitch their technology to engage investors and industry partners. While James alone could describe his science, he couldn't see how to find the signal of authentic Demand in the noise around him, but James plus Ben was a whole different ball game. Together, they had complementary capabilities. They could hold the science and probe what the world actually wanted together.

This shift matters more than it might appear. There's an old proverb, African in origin, that captures something learned through hard experience: if you want to go fast, go alone; if you want to go far, go together. There are few situations that saying holds more true than when spinning out deep technology from an institute.

Translation is a team sport.

Our time together helped James and Ben to look at what they actually had, not as individuals, but as a unit, and to see how far they could really go, together. It turned out that together they had precisely what they needed to start to cross the bridge. Deep expertise in a specific area of biology that others had overlooked. A mechanism they believed could work where others had given up. Early data suggesting they might be right. A network from their years at a major research institute. Each other. And the combined time they were willing to commit.

Testing the forces

When James and Ben sat down to work through the forces of Supply, Demand and Capital for the first time, the exercise seemed straightforward enough. Articulate what you know. Name what you don't. Find the gaps. Fill the gaps.

James started with Supply. This part came easily to him, having developed the research over many years. He could articulate what they knew about the biology and the insight they held that others had missed. He could describe what they were – two researchers with complementary expertise. And he could list what they had – early data, equipment access and time to commit. The Supply picture filled quickly, dense with assertions they believed they could defend.

But when he moved to Demand, the pen stopped. Who was actively searching for what they were building? He had a sense that others struggled with these kinds of targets. He'd heard complaints at conferences and he'd read about failed programmes. But when he tried to write something specific, something he could point to as evidence that someone out there was looking, he found he had nothing. He had no names and he'd had no conversations. He didn't know of any real Demand – he only had vague notions and soft assumptions.

So he wrote a question mark and moved on, knowing he had work to do, but also knowing he now had someone with him to help accomplish it.

Then he turned to Capital, and it's here that the challenge became impossible. He knew what he needed, which was funding, infrastructure and knowhow. But this is only part of the Capital challenge.

He also needed to work out what those with resources to commit wanted in return. And he had never thought about the return. He had only ever thought about sources of Capital the way Kusal once had — as something you extracted from the system to continue to do good work, not as an exchange where both sides had to see value.

By the time he was ready to make decisions about how to move forward, the imbalance was obvious. He had Supply down pat but didn't understand Demand — what was he building, and for whom? How would they deliver it? And when looking for Capital, what would Capital providers need from the relationship?

James looked at the gaps in his knowledge and now understood why he'd never been able to answer these questions. It was because the case for answering them wasn't something he could assume. It was something that emerged only when he started testing for authentic Demand. It was this step that uncovered the forces at play on James and his work, and how he needed to manage them to move his project forward.

James realised he couldn't answer these questions because they depended on an understanding of forces that was only one-third complete. And this also explained why he was busy, but getting nowhere — because what he was building kept shifting since he didn't know who he was looking for, or who might be searching for him.

James was getting the picture, but it was Ben who took a step back and said what they were both thinking.

'We don't actually know if anyone wants this.'

That realisation was not only confronting, it was also unsettling. But they realised something else. The gaps in their understanding weren't failures. They were a map of the discovery work they needed to do in order to find the answers they wanted.

Testing, testing

What followed was six months of experiments, but not in the lab. These happened in the world.

Ben and James started testing for Demand. They asked, who needed what they had? How could they deliver it? What was the approach? They thought about what they knew, and Ben took the guesses they'd made about who might be searching and started checking them against reality. He had conversations with people who'd spent careers in the space they were trying to break into. He met with executives who'd heard a hundred pitches that month, investors who could spot wishful thinking from across a conference room and former colleagues who'd crossed over from research to industry and could tell them what actually mattered on the other side.

Each conversation was a test. Did their assumptions hold, or did they break?

Many broke.

They'd assumed the targets they were pursuing were active problems across their industry. But when they tested this, they discovered that several major players had abandoned those targets years earlier. They'd run lab experiments and hit walls, and so they moved on. The

110

Demand James had assumed was there had evaporated long before he'd even started looking.

But those same conversations revealed something else. There was a subset of organisations that hadn't given up. They were still actively searching for new approaches because even though their existing programmes had stalled, they could see the commercial opportunity. This was Demand that James and Ben hadn't expected because it wasn't where they'd assumed it would be. It was somewhere else entirely.

They'd also assumed a broad approach would be attractive, offering flexibility to pursue multiple directions. But in testing, they discovered that would-be investors wanted specific programmes with focused milestones, not possibilities that could go anywhere. They went back to their mapping and found that their product thinking, which had been a mess of competing choices, started to resolve into a single, strong option.

They'd assumed their two-person team was sufficient to get started. But once they started looking for authentic Demand, they discovered that serious backers expected operational capability they didn't have. They realised they needed someone who understood how to build an organisation rather than just run experiments. The exercise had revealed an execution gap they couldn't fill themselves. So they filled that gap too, with Anna.

Anna had spent years in roles that required translating scientific promise into propositions investors could evaluate. She'd occupied the vantage point James and Ben were trying to reach. When she joined, she brought not just experience but a different way of seeing, a complement to the science that neither of them could have provided alone, and she saw immediately what they couldn't see about their own position. Anna joining the team was a direct response to a gap the process had made visible, one they could now address because they finally knew it existed.

Narrowing their approach and expanding their team were both answers to gaps they found by testing Demand. And as they tested, and retested, they continued to find gaps that needed filling and insights that guided their next step. Each gap improved their understanding. Each insight was a step toward a foundation that could bear weight.

They also discovered what already worked. They discovered that their core insight, the thing that made their approach different, resonated with people who understood the problem. They discovered which of their evidence would move a conversation from polite interest to genuine intent. And they discovered what would need to be true before anyone would actually commit to the project. And with each test, the picture sharpened into something they couldn't have learned any other way.

By the time the company spun out in 2023, it bore little resemblance to how they thought it would look. What they had was no longer a broad description of possibility, but a grounded picture of what actually set them apart. They knew who was interested because they had spoken to them. They understood what they were asking for because they knew what they were offering in return. What they were building was no longer a list of options, but a clear product direction. From within that sea of Demand, the market was defined. The work of execution belonged to a team, not just two individuals.

And the case they were making had earned itself. The idea that had taken shape in a garage and drifted for two years without direction had become something that could bear the weight of serious scrutiny and attract Capital with it.

Three forces, fully visible

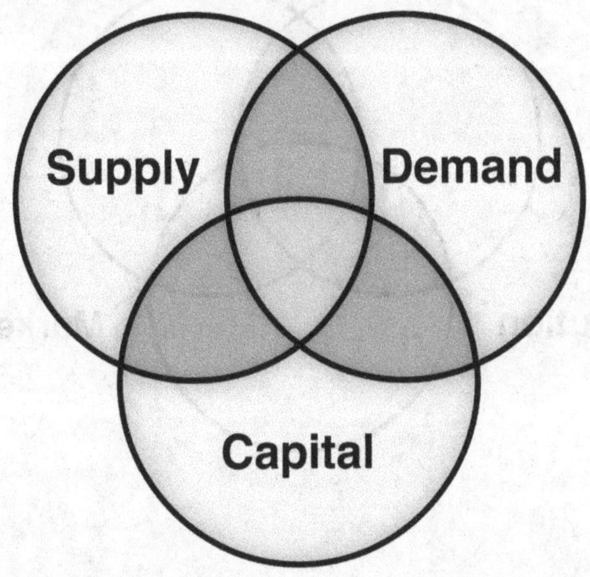

Figure: The three forces you can read: Supply, Demand and Capital

James's journey with Demand completes the picture that Marcus's began with Supply and Kusal's extended with Capital. Together, they show us that these three forces – Supply, Demand and Capital – are always operating on and around you and your work. You can't control them directly, but you can learn to read them.

And once you can read them, you can learn to pull three specific levers that will align the beliefs you hold about your work with the opportunities these forces are showing you. That's where we're going next.

113

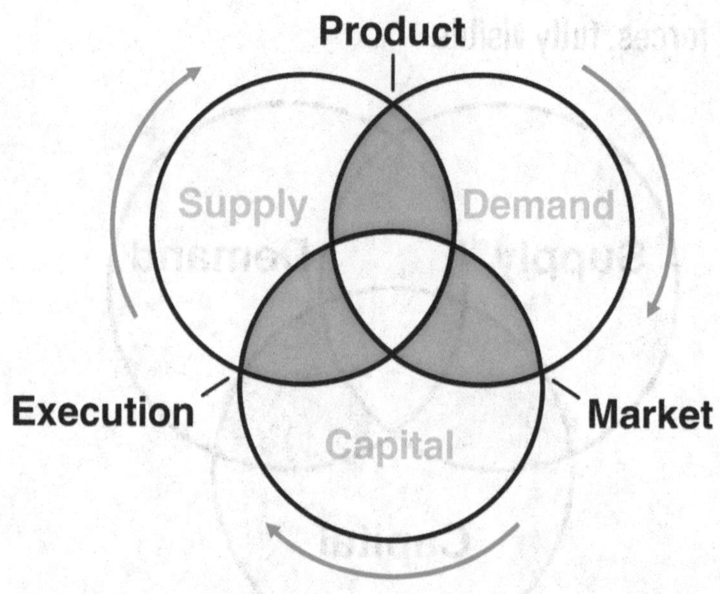

Understanding the forces tells you what you're up against. It doesn't tell you what to do. There are levers that, when pulled, create movement and provide direction. These levers turn Supply into a Product the world can use, that find the Market within the noise of Demand, that make the promise of Execution credible enough for Capital to flow. Most researchers never find them. But what if the difference between a career spent waiting and one spent moving was knowing which levers to pull – and how?

Chapter 7

PULLING THE LEVERS OF PROGRESS

Sadhika had a good foundation in her reading of these forces, or thought she did. Four years of doctoral work had produced a detection system that could identify skin cancer without cutting into the patient and the pilot study showed 94% accuracy. Plus, it worked across all skin colours, a huge limitation of the current systems that biased fair skin. She wrote the publications, completed her PhD and seemed to have the whole apparatus of academic success firmly in place.

But she still had no idea what to do next.

She had options. The technology could become a diagnostic tool for clinics. It could be licensed to an existing company. It could be spun out as a startup. It could even return to the lab for further development. Each path had advocates, and each carried risks she couldn't fully see. But nothing in four years of doctoral training, in her research, within her expertise as built by the traditional system, had prepared her to choose between them, and so she was paralysed, which is what happens when you've proven something works but don't know what that proof actually means.

Foundations like this that haven't been tested against the world feel solid until you try to build on them. Only then do you discover you don't know which parts can carry the weight of what you want to build and which will crumble the moment pressure is applied. Sadhika had a technology, she had results and she had credentials. But she had never tested whether the solution she had in her head – what she had built, who it was for and how it would be used – would survive contact with reality.

The permission most researchers never get given

Sadhika was based in Australia, where skin cancer is not an abstract concern. Two in three Australians will be diagnosed. One person dies every six hours.[38] Early detection, catching a melanoma before it spreads, is the difference between a minor procedure and a death sentence. The clinical need for better detection is an ongoing Demand force. Sadhika's technology addressed a real problem.

Yet, the path forward was not obvious at all.

Fortunately for Sadhika and her research, her supervisor understood something that most supervisors miss.

In most laboratories, translation is a distraction at best and a career risk at worst. A doctoral student who spends time on commercial conversations instead of publications is a doctoral student falling behind, and the incentives are clear enough in universities and institutions that most supervisors enforce them. They don't do this out of malice, they do it because the old system punishes anything else and they're trying to protect themselves and the people they're responsible

---

38   Cancer Council Australia. (2023). 'Skin Cancer Statistics and Issues.' Retrieved February 18, 2026 from https://www.cancer.org.au/cancer-information/types-of-cancer/skin-cancer.

for. The old system's metrics offer no credit for Market exploration, no recognition for conversations with industry, no reward for the work of discovering whether anyone actually wants what you've built. Time spent on translation is time not spent on papers, and papers are what the old system counts.

Sadhika's supervisor had decided, years earlier, that the system was wrong about this. He had watched too many promising technologies die in the gap between discovery and deployment. He'd seen too many doctoral students complete rigorous work that never reached the patients it could have helped. And he had concluded that something in the standard model was broken. In truth, he recognised that the old system simply wasn't working. And he came to believe that translation wasn't a diversion from research but a different form of it, a different kind of question requiring a different kind of experiment. So he built his laboratory around the conviction that both kinds of work deserved protection.

In a system that often punishes researchers for looking outward, having a supervisor who saw translation as legitimate work gave Sadhika something most researchers never receive: permission to explore without apology. Not grudging tolerance or a blind eye turned while she pursued something the supervisor considered peripheral, but active encouragement to discover what her technology might become in the world beyond the university. She could talk to clinicians without worrying that she was neglecting her real work. She could meet with companies without fearing that her supervisor would question her priorities. She could test her assumptions against reality without the guilt that paralyses most researchers who try to do the same thing.

Not every researcher has this. The ones who don't face a harder path, having to grant themselves the permission that no one else will give them. Sadhika was fortunate, and she knew it, and she used the space she'd been given to great effect.

What can I do today without needing anyone's permission?

We began by focusing on how to take the assumptions she had made about what she had and turned them into assertions she could test. This is how she would learn what she really had to work with and how to truly read reality.

This approach has a name. Entrepreneurship scholars call the underlying logic 'effectuation', a way of reasoning that expert entrepreneurs actually use when they need to make progress under genuine uncertainty.[39]

Traditional planning asks what you want to achieve and then works backwards to determine what resources you need. It assumes you can define the goal clearly, identify the gap between where you are and where you want to be and then acquire whatever is missing. This works well enough when the destination is known, the resources are readily available and the path is reasonably clear. In other words, it really works for problems that aren't actually uncertain at all.

Effectuation inverts this entirely. Instead of starting with what you want to achieve, it starts by asking what you have and works forwards to discover what's possible. The logic rests on the process of testing

---

39 Sarasvathy, S.D. (2001). 'Causation and Effectuation: Toward a Theoretical Shift from Economic Inevitability to Entrepreneurial Contingency.' *Academy of Management Review*.

assertions, refining those assertions and then retesting them until you find the answer to the question that changes everything: *what can I do today without needing anyone's permission?*

The distinction matters because it changes what counts as a valid starting point. You don't need everything to begin, and you don't need to see the whole path. You need to know what you actually have, stated honestly, and then you need to move.

Finding what she had

Sadhika started with what seemed like the obvious question – *what did she actually have?*

Her PhD work had produced two distinct assets. The hardware – a specialised camera system – was entangled with her university, which meant there was a thicket of existing IP agreements that would take lawyers months to untangle. The other asset was the software – the algorithm that interpreted the images and flagged potential cancers. This was cleaner, more clearly hers, but also invisible – just code running on a machine. That made it more abstract, which made it harder to demonstrate to someone who wanted to see a thing rather than hear about a capability.

So from the start, Sadhika had been thinking of her Product as a device. A new type of camera that happened to have intelligent software attached. But framing her work around what's easiest to show isn't the same as framing it around what the world is demanding. And she had never tested that framing against the people who would actually use her work: her Market.

So she arranged to observe how skin cancer screening happened in practice. Not in theory, not in the literature, not in the sanitised descriptions that appear in grant applications, but in a busy dermatology clinic with patients cycling through and real constraints pressing on every decision.

What she found was that the rooms were small. Every surface was occupied. Dermatoscopes and imaging systems and computers and examination tables were all necessary, and were all competing for the same limited space. Equipment fatigue was visible everywhere she looked. There were just too many devices from too many vendors, too many systems that didn't talk to each other and too many promises of integration that had never been kept.

The dermatologists she spoke with were polite but direct. They didn't want more hardware. They were drowning in hardware. New hardware meant new training, new maintenance contracts, new workflows to learn and new space they simply didn't have. They were firm in their stance: anything new would need to work with what they already owned. And anything that Sadhika brought to them needed to be something that would make their existing investment smarter, rather than demanding they make a new one.

In addition to the IP issues surrounding the camera, it also brought with it manufacturing complexity and required physical space that the clinicians simply didn't have. If she stuck with her current iteration, she would have problems that would take years to solve, problems that had nothing to do with whether her technology actually worked and everything to do with whether anyone could adopt it.

But her algorithm had none of those problems. Its invisibility became its value. In theory, her software could run on equipment clinicians already had. It could integrate into workflows that already existed. It could update remotely, improve continuously and scale without shipping physical units across the country.

Sadhika had walked into that clinic convinced her camera – the equipment itself – was the Product. But when she tested that belief against the reality of a working clinic, the belief broke. And when she walked out she understood that the camera was actually the liability. It was her algorithm and know-how that was the real Product.

Turning uncertainty into certainty

What Sadhika was doing, without quite having the language for it yet, was running a loop that systematically turns uncertainty into certainty. And the loop begins with what you can control.

What you can control are the levers you shape and pull in response to what you see around you: Product, Market and Execution. The Product you shape from your Supply. The Market you identify from within the broader force of Demand, specifically enough that you could reach it tomorrow. And the Execution capability you assemble through the deployment of Capital to deliver on what you promise.

These are the levers that generate progress, and you won't find momentum until they align. That alignment cannot be assumed. It must be tested.

Testing for alignment

In the testing undertaken by Marcus, Kusal and James, we looked at components of their research and the landscape it sat within – was there Demand? Who controlled Capital? What return would they expect? We were testing to understand what forces were shaping the terrain that the research and researcher stood upon. Supply, Demand and Capital set the context.

But now that the shape of Sadhika's Supply was clearer, she could see Demand and even had ideas of where Capital might flow from, so the question changed. It was no longer just *where's the Demand?* She needed to know whether her understanding of what she had would actually survive in the real world. She needed to understand whether she had a Market for her Product.

This is where most researchers go astray. They operate on assumptions about their Product, their Market and their ability to Execute. But assumptions blind us to reality. Assumptions are beliefs you carry, without proof or certainty. Assumptions might feel safe because they aren't challenging how you see the world. They sit quietly shaping your decisions without you having to test whether they're actually true or not. But untested assumptions are poison pills in translation, because they can lead you to build an entire project on something that was never true.

What we need to be able to truly act, is an assertion. An assertion is the belief behind an assumption but articulated to make it explicit and testable.

Rather than: *This is true.*

It becomes: *I believe this to be true, and can test it by...*

Once restated clearly, an assumption becomes an assertion, and it now becomes knowable and testable. And when you take that assertion out of the lab and expose it to reality, you can find out whether it holds. If it does, you've strengthened your position. If it doesn't, you refine it, reshape your levers and run the loop again. Either way, you've reduced uncertainty.

The LARC Loop

This rhythm of testing assertions has a shape: Learn, Assert, Reality Check. And then repeat. This is the LARC Loop. From what you have learned to be true, you assert something concrete about your Product, your Market or your Execution capability. You then reality check it with a structured real-world test. And you learn again, from a stronger position.

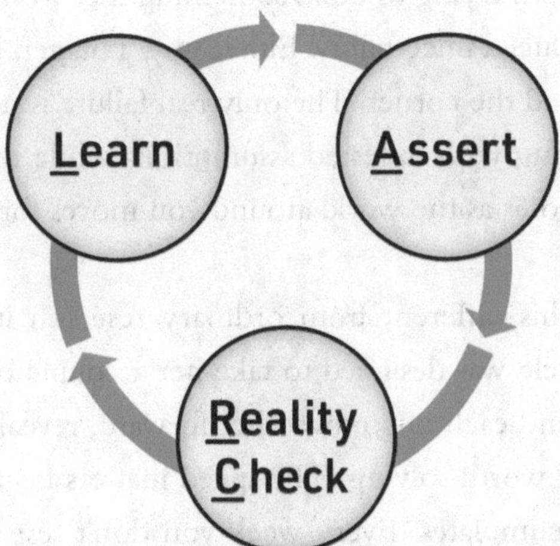

Figure: The LARC Loop. Each cycle reduces uncertainty and builds on the previous. The loop continues as long as the work does.

You aren't pulling levers randomly. You're shaping Product, refining Market and strengthening Execution, cycle after cycle, harnessing the forces around you and building alignment to drive translation. This is how uncertainty becomes certainty.

Sadhika was testing her work through a series of LARC Loops. She learned from what she had, the imaging techniques and algorithm and pilot data she'd accumulated through years of work. She asserted something testable about how that capability connected to the world, in her case that the camera was the Product. She reality checked that assertion against people whose own work depended on the answer, and it broke. The breaking taught her something she couldn't have learned any other way. Then she used this new learning to run another LARC Loop, from a stronger position than before, and reality checked that too.

The LARC Loop is valuable because it either confirms what you believe, which means you can build on it, or it breaks what you believe, which means you've discovered something that isn't true and prevents you from trying to build something that won't work. Either way, you've reduced uncertainty. Either way, a bigger, better problem is waiting around the corner. The only real failure is not running the loop at all, sitting with untested assumptions while the window for testing them closes as the world around you moves forward.

What makes this different from ordinary research iteration is the speed. Each cycle was designed to take her a couple of weeks rather than months, and each one moved her forward, revealing a new and better problem worth solving. The speed matters because otherwise uncertainty accumulates. Every week you don't test an assumption is a week that assumption might be wrong without your knowing it. Every month you spend building on an untested foundation is a month you might be building in the wrong direction entirely.

Eric Ries, whose work on lean operating methodologies shaped a generation of startups, put it simply: the company that wins is the one that iterates fast enough before it runs out of resources.[40] The same logic applies to research translation. The researchers who succeed are the ones who test fast enough to stay ahead, who would rather discover they're wrong early than discover it after years of misdirected effort.

This is the mindset that the new system rewards. Yet this is also the opposite of what traditional research training expects.

Doctoral programmes teach you to wait for certainty before making claims, to qualify everything, to hedge against being wrong. The instinct makes sense, as premature claims can undermine the credibility you've spent years building. But in translation, the greater risk is moving too slowly, carrying assumptions for years that a few well-designed, well-executed reality checks could test in days.

The loop teaches you to trade uncertainty for insight as fast as you responsibly can. Hold a strong view, but hold it loosely. Make assertions with conviction, then actively try to prove yourself wrong, because it's far better to be proven wrong in a quick test than after you've invested years and millions of dollars in a direction that was never going to work.

---

40   Ries, E. (2011). *The Lean Startup: How Today's Entrepreneurs Use Continuous Innovation to Create Radically Successful Businesses.* Crown Business.

One researcher put it simply to me early on: 'I wish we'd done this seven years ago. Why didn't someone tell us this was never going to work?' Now you know why: the old system never rewarded those who asked this question. Times have changed.

Sadhika's first assumption – the camera is the Product – was so deeply held she hadn't recognised it as an assumption at all. It was only when she reframed it as an assertion – I believe the camera is the Product – that things started to move for her. That's because when she named it and reality checked it in a real-world test, the assertion broke. Only then could she revise her learning. The new assertion – I believe the algorithm is the Product – emerged from what she'd learned.

This revision wasn't a setback – it was progress, because now she had a belief that was actually shaped by reality, rather than shaped by what she'd happened to construct during her PhD.

The revised Product assertion needed a revised test to find a Market. If the algorithm was the Product, who would want it enough to commit Capital? Who was already searching for better detection capability, already had cameras deployed, already had relationships with clinics and already had infrastructure that her software could plug into?

She found the answer in the technical team for a network of skin cancer clinics – they fit every criterion. They were operating a network of clinics across the country. Their dermatologists trusted their equipment. And they were actively searching for ways to make their operations smarter and to differentiate their offerings in a competitive Market. They weren't waiting to be convinced that better detection mattered. They were already looking.

Sadhika had been wondering who she might persuade to want her technology. But it turned out, someone was already looking for it.

But interest isn't intent. Interest is a curiosity, a willingness to keep talking. It's a door that hasn't been closed… yet, but still could be. On the other hand, intent is the willingness to commit resources, time or access. It's the willingness to actually do something rather than just discuss the possibility of doing something.[41] This company was interested. The question was whether they had any intent to invest.

So Sadhika proposed a test. Let her run the algorithm on their existing image database. Let her demonstrate what it could see that their current tools couldn't. She didn't suggest any commitment beyond the data access she'd need to run the test and a chance to prove the concept. If it worked, they could discuss what came next.

They said yes.

For Sadhika, her disciplined adoption of the LARC Loop drove her project forward, fast. She saw that a partnership with a data-sharing agreement was promising, but it was not, by itself, a path to adoption. She had asserted that a company would partner with her if she could demonstrate improved detection. She had tested it with a concrete proposal. The assertion held. So now she knew that Demand from a specific Market was real, not theoretical, because there was a company willing to share its data and explore where this might lead.

---

41 The progression from interest to intent to investment draws on the customer development methodology pioneered by Steve Blank. See Blank, S. (2005). *The Four Steps to the Epiphany: Successful Strategies for Products that Win*. K&S Ranch.

She had moved from interest to intent, which meant she had a new problem to solve. But it was a bigger and better one than the problem she'd just put behind her. How was she going to get this closer to the clinic? She was going to need to run a clinical trial.

Accumulated learning as a path to translation

To run a proper clinical trial, the kind that would generate evidence for regulatory approval and demonstrate real-world effectiveness, Sadhika needed more than a willing partner. She needed funding, infrastructure and time protected from other demands. However, she also needed credibility with funders who had seen too many promising technologies fail in translation and had learned to be sceptical of potential that hadn't been tested against reality.

Her supervisor continued to provide cover, treating her commercial exploration as legitimate research development rather than career distraction. She applied for a newly-created fellowship designed for exactly this purpose, funding researcher-entrepreneurs as they attempted the crossing from discovery to deployment.

When she applied, she wasn't presenting untested potential. She was presenting accumulated learning that she could demonstrate. She had a focused Product based on her own Supply that she could execute, a validated Market in the form of skin cancer clinic IT teams and an Execution pathway to deliver clinical evidence in the form of the clinical trial with access and infrastructure provided by her industry partner.

She had done the work – the application showed it. And she received the fellowship, giving her the initial Capital she needed to accelerate Execution.

The partnering company deepened its commitment in response. Not just data access now, but co-investment in the clinical trial itself. Intent had become investment, which meant Sadhika now had a bigger problem once again, and a better one too: could this technology scale?

The trial was about to begin, and with it the testing of a whole new set of assertions. Would the accuracy hold in real-world conditions, across different clinic environments, with different operators and different patient populations? Would clinicians trust the algorithm's recommendations enough to act on them, rather than defaulting to the tools they already knew? Would the economics of deployment work at a scale that justified what everyone had committed? Each question was larger than the last, which is what progress looks like when the LARC Loop is working: not fewer problems, but better ones, the kind that only emerge because you've solved the problems that came before them.

Something is building

The loops don't stop. The rhythm doesn't end. It's not a sequence you complete once and move past, it's a cycle you sustain for as long as the work continues. The researchers who successfully translate are the ones who maintain it, holding their beliefs firmly enough to act on them and loosely enough to abandon them when evidence demands.

Each cycle doesn't just reduce uncertainty. It builds something. It reshapes your Product, sharpens your Market and strengthens your Execution. You're not just testing ideas – you're refining the three levers you control.

The insight Sadhika gained from visiting that clinic didn't disappear once she'd revised her Product assertions. It became part of what she knew, part of what made her credible, part of what differentiated her from anyone who might try to follow the same path. The relationship with the clinics didn't evaporate once the data-sharing agreement was signed. It deepened, creating commitment that would be hard for a competitor to replicate, even if they could see exactly what she'd done. The fellowship didn't just provide funding. It signalled validation that would open doors she didn't yet know existed.

But there's something else happening in this rhythm, something the loop alone doesn't quite capture. Every loop was building more than learning. Sadhika was building something in the truest sense, though she couldn't yet see its shape. Each loop added to it. And whatever it was, it was starting to feel like it mattered more than the science itself.

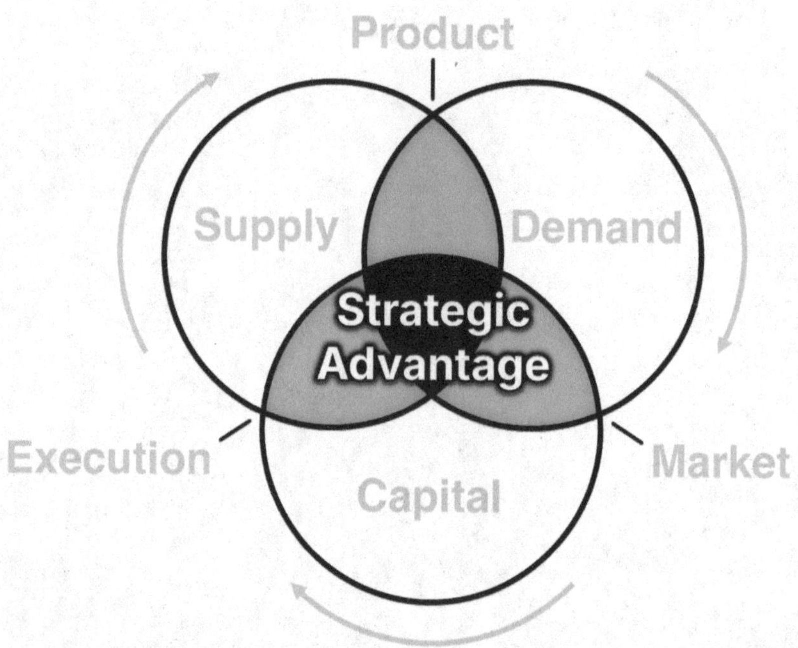

Excellence gets you to the table. But every researcher you've met in these pages had competitors with comparable science and expertise. What keeps you ahead is something else entirely, something earned rather than claimed. And the real question becomes: what Strategic Advantage are you building every time you make a move?

Chapter 8
STRATEGIC ADVANTAGE IS EARNED

Marcus wasn't the only scientist who understood inflammatory pathways. There were laboratories around the world pursuing similar mechanisms, researchers with comparable expertise, teams that could trace the same biological logic to the same therapeutic possibilities.

James and Ben weren't the only scientists exploring protein degradation. The targets they were pursuing had attracted attention precisely because others had seen the same opportunity.

Sadhika wasn't the only researcher developing detection algorithms. The problem she was solving was obvious enough that dozens of teams were working on variations of the same approach.

And Kusal wasn't the only researcher pioneering spatial biology. The techniques he'd developed were emerging in laboratories on three continents, and the window in which he was one of the only ones building them wouldn't stay open forever.

The science was comparable. The expertise was comparable. The opportunity was visible to anyone paying attention.

So, why these researchers? Why these projects?

Necessary, but not sufficient

The traditional system has an answer to why some researchers succeed where others don't, and it's the answer that researchers absorb from the first day of graduate school.

Excellence.

In other words, if you build the strongest scientific case, success will follow.

Yet you've likely known for years that there's something missing from this equation. The paradoxes were everywhere: Marcus had global recognition and hundreds of publications, yet his science was marooned in the laboratory going nowhere; James and Ben's unique insight into abandoned targets had left them unable to find a direction; Sadhika's superior software led only to paralysis; and Kusal's pioneering excellence translated only to failing computers.

Every one of these researchers was excellent. But so were the researchers working on the same problems who never translated at all, the ones with comparable expertise and comparable science who are still in their laboratories wondering why nothing's moved.

The reason is that excellence doesn't compound. You can become more excellent, certainly, but that's not the same as building something that grows. Excellence is a threshold you cross, not momentum you build. And to achieve translation, you need momentum.

The flywheel

Physicists have a name for an isolated system where early effort creates lasting momentum over time. They call this 'conservation of momentum' – and one of the best physical examples of this is the flywheel.

A flywheel is a heavy disc that resists changes in rotational speed – both acceleration and deceleration. It takes enormous energy to get it spinning, and the first rotations are the hardest, requiring force that seems disproportionate to the movement you achieve. But once the flywheel is moving, something changes. It stores the energy you've put into it and each subsequent push adds to what's already there, building on the momentum that your previous effort already created. Eventually, the wheel spins faster than any single push could achieve, carrying an accumulated force that would be impossible to generate from a standing start.

This is the dynamic that's been unfolding across the previous four chapters, even though we hadn't named it yet. Each disciplined adjustment to Product, Market and Execution was another push on the wheel.

Marcus's flywheel started spinning when he walked into that first investor meeting, the one he'd been certain would never happen, that his assumptions had told him was impossible. That conversation didn't just provide validation. It provided a reframe. Lung disease rather than kidney disease, a direction he hadn't seen, revealed by people who were scanning the landscape from a different vantage point. In a single conversation, the Product he was building and the Market he was building it for had both changed.

Kusal's flywheel started spinning at the conference where his reframed narrative revealed something he hadn't fully recognised himself: a Product that the world could see. And the response he received revealed a Market of people searching for exactly what his platform could do.

James and Ben's flywheel started spinning when months of Market testing broke their assumptions, and subsequently revealed a subset of organisations that hadn't given up on the targets they were pursuing. They found Demand they hadn't expected because it wasn't where they'd assumed it would be. Once they could see this Market clearly, the Product they were building and the Execution it required started to take a different shape entirely.

Sadhika's flywheel started spinning in a dermatology clinic where she discovered that the Product she thought she was building wasn't what the world would use. The revised Product attracted a partner who had been searching for the exact capability she'd developed, which gave her both a Market and an Execution pathway to reach it.

Each of these researchers found their first rotation of the flywheel in the same place: a fast and disciplined reality check with someone whose own future was impacted by the answer. Not a colleague offering an opinion or an advisor suggesting a direction, but someone with skin in the same game, whose response carried weight because something real was at stake.

Those first rotations of the flywheel were the hardest, requiring force that seemed disproportionate to the movement it achieved. But once the wheel was turning, each subsequent push added more momentum to what was already there, until it felt like it could almost move entirely on its own.

Bill Gross, who has founded or backed over 200 companies through his firm Idealab, spent years trying to isolate what actually separated success from failure. He had enough data to look for patterns that anecdotes couldn't reveal, and he analysed them systematically, trying to identify the factors that mattered most.

What he found surprised him. The single biggest factor wasn't the idea. It wasn't the brilliance of the innovation, the elegance of the solution or the quality of the science. Such excellence accounted for only 28% of the difference between success and failure. Important, certainly. But not decisive.

And it wasn't the team, the talent, the experience or the Execution capability either. These accounted for 32%. This was more important than the idea, but it still wasn't the largest factor for success.

Instead, the biggest factor was timing. Whether the world was ready. Whether authentic Demand had produced a Market ready to accept the innovation. Timing alone accounted for 42% of the difference between success and failure.[42]

Researchers will recognise what this means. The traditional system trains you to believe that if the science is good enough, success will follow. Gross's data says otherwise. Excellence matters, but the team matters more. And timing matters even more. The question isn't just whether your science is strong. It's whether you have the ability to Execute and, most importantly, whether the world is ready to receive it.

---

42 Gross, B. (2015). 'The single biggest reason why start-ups succeed.' TED Talk. Gross analysed over two hundred companies and found that timing accounted for forty-two percent of the difference between success and failure.

But here is what the data doesn't tell you, and what our five researchers discovered: timing isn't something you wait for. Timing is something you discover, and even develop, through motion. Through taking action.

None of the researchers we've met in these pages waited for the timing to be right. They kept moving until they found where it was right. And once the flywheel began to spin, timing didn't precede momentum – it accelerated it. Each rotation revealed new opportunities, compounding into something a competitor starting today would need years to match.

Building your Strategic Advantage

The forces aren't just impacting you alone – they are shared. Supply, Demand and Capital look similar to every researcher in the same field, who knows to look for them. A dozen teams might notice that a particular area attracts investor interest. A dozen researchers might recognise that a certain technology is gaining momentum. Once you learn to read the landscape, the opportunities become visible.

While you may gain a small advantage from being better at reading these forces, there's no lasting advantage from a shared landscape. Advantage comes from what you do with this knowledge.

Each time you test an assertion and it holds, you've learned something a competitor hasn't. Each time you act on that learning by shaping your Product, engaging a Market or strengthening your Execution, you create something harder to replicate. Not because the forces you read are unique, but because what you decide to build in response is.

Over time, those decisions accumulate. You build relationships with people who trust you because you've shown up and delivered. You gather evidence about what works and what doesn't. You develop capabilities – technical, commercial and operational – that make the next step easier than the last. None of these things appear overnight, but together they begin to change your position.

You are learning to pull on those internal levers to harness those external forces and turn them into an aligned and unique opportunity that you alone can execute. This is what the flywheel builds at the centre of it all, the unique position that emerges when you can leverage the landscape with deliberate action. We call it Strategic Advantage.

Strategic Advantage isn't excellence, though it requires excellence as its foundation. It isn't luck, though timing plays its part. It's the position you continue to build through motion, the relationships and evidence and capability and earned trust that compound with every rotation of the flywheel until a competitor starting today couldn't replicate it, even if they could trace every step you'd taken.

Every researcher in these chapters was building it, whether they knew it or not. And by the time anyone else noticed what they were pursuing, they had built something that couldn't be caught up to.

Where those flywheels carried them, and what the crossing did to them in the process, is what the next part of this book is about.

The Complete Breakthrough Model

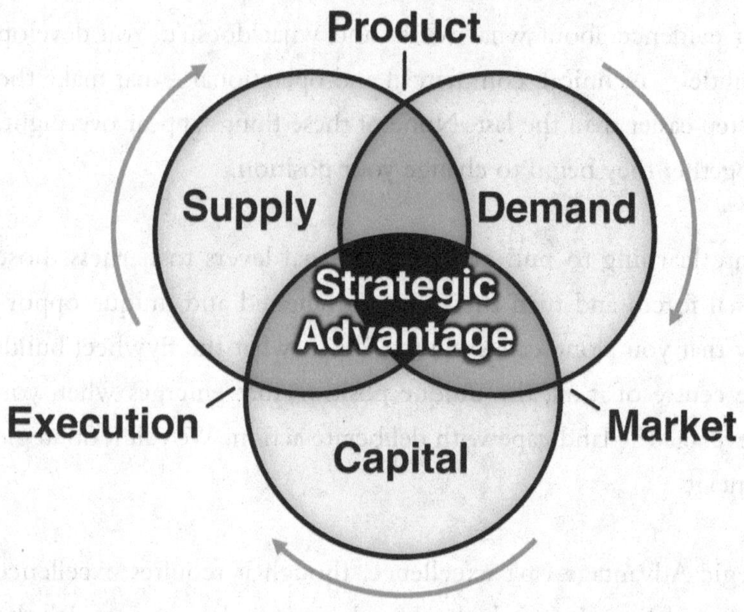

Figure: The Complete Breakthrough Model. External forces, aligned by internal levers, systematically building Strategic Advantage.

Now that we've covered Strategic Advantage the *Breakthrough* model is complete. And with the complete model you can now see the whole picture. You know that the forces of Supply, Demand and Capital are always moving around you, pressing onto your work, and shaping what's possible whether you can see them or not.

You also know why and how to pull the levers of Product, Market and Execution, as the elements that you can actually act upon. You know that when you learn to read the forces clearly by taking assumptions, turning them into assertions and testing those assertions, you can learn to pull those levers with discipline to create powerful results.

Now your work can stop feeling like a series of disconnected efforts and become a coordinated process for generating real momentum towards impact.

This is Strategic Advantage, which you build through repeated cycles of learning, testing and refining. Each reality check strengthens what you know, what you build and how you bring it to the world. Over time that alignment compounds, and you stop waiting for the world to recognise your work and start building something the world can respond to.

BREAK-THROUGH

This book has covered the breakdown of the old system, the emergence of the new one and the pathways, forces and levers now driving translation.

Up to this point we've looked at how the research system has traditionally worked, where it's failing and how researchers might begin to navigate it differently. We've looked at the forces that shape what moves and what stalls. And we've seen the levers that take assumptions and turn them into assertions we can act on to build our own Strategic Advantage towards impact.

Part III shows us what the new system provides researchers who are willing to step out onto the bridge and make the crossing. It shows what becomes possible when the forces are no longer invisible, the levers no longer abstract and deliberate action replaces passive waiting and hoping.

More importantly, it reveals what research – and the life of a researcher – looks like on the other side of the bridge.

While every path across the bridge is different, the pattern is not. These researchers show you what's possible for you too.

Chapter 9

THE CROSSING

We left Sarah in Chapter 1 at the moment the mathematics finally made sense. The moment when she understood that the contract she'd unwittingly agreed to had been broken by forces that had nothing to do with her. The relief had been physical, she said. But once the relief settled, she was still standing in the same place, in the same lab, doing the same excellent science that still wasn't reaching the people it was meant to help.

This chapter is about the moment a researcher stopped waiting to be recognised and started positioning their work in a way the world could engage with and act on.

The first step over the new bridge

A conference was approaching, one where the audience would include clinicians, industry representatives and patient advocates alongside the usual researchers. Sarah was asked to present her work. This wasn't unusual. She'd given hundreds of talks by then and knew how to construct them the way researchers are trained to construct them: hypothesis first to establish intellectual foundation, methodology next to demonstrate rigour, findings to show what you've contributed and implications to prove you're thinking beyond the immediate study.

This sequence is well known, and because of that it earns the speaker credibility from the listener, proves you belong and justifies your right to be taken seriously. But Sarah had accumulated years of polite applause and adequate questions. What she hadn't accumulated was any sense that her work was connecting with anything or anyone beyond the closed loop of publications and citations and grants. So, this time, she decided to try something different.

Instead of following the set sequence, she opened her talk with the human side of her research. She talked about the lived experience of patients who had been let down by the current standard of care. This naturally led to a discussion about the problem – not a research problem framed in the language of hypotheses and mechanisms, but a human problem described in terms of what was being lost because no one had solved it yet.

Only after covering that human position did she turn to her research, positioning it as one possible response to something that mattered, rather than as the point of the talk itself.

The new sequence to her presentation felt intellectually wrong. But it felt emotionally right. And when she finished, something happened that had never happened before. People lined up to talk to her. Not just polite 'thank yous' or 'that was interesting', but actual conversations about things that had the potential to take her from being stuck to having real momentum.

These were people who had seen something they wanted to understand better, who had work they thought she might help with, who hadn't cared so much about her methods but cared deeply about the problem she was trying to solve. And these same people could see and feel that she cared about it too.

Sarah hadn't changed her science. She'd changed how she talked about it. And in that small adjustment, putting the human experience first, she came to understand something larger than a new communication technique. She was able to show the world *why* her research mattered – and open up conversations with those who believed it mattered too.

From producer to partner

Sarah's presentation was only the first step across a bridge that was waiting the whole time, but that she hadn't been trained to see.

As we've seen, researchers are trained to think of themselves as simply producers of knowledge. You generate findings and publish them. What happens after that is someone else's job.

This wasn't Sarah's nature, nor what had drawn her into research, but she had absorbed this so completely it had become invisible. Her talks led with research because research was what she produced. Her value was now measured in publications and grants because those were the metrics that mattered to the system she operated in. Her identity was that of a researcher, and researchers research.

But when she led with the problem instead of the solution, when she described the people struggling rather than the mechanism she'd discovered, something shifted in how the audience responded. They weren't evaluating her credentials anymore. They weren't even really looking at her research. They were recognising a shared problem and seeing someone committed to solving it. And that changed the way the world saw what she had to offer.

The insight seems obvious in retrospect: the world doesn't want your research. The world wants solutions to its problems, and your research is one possible way of getting there.

As a result, Sarah had spent years experiencing industry relationships as most researchers do: as one-directional. She asked; they decided. She needed things from them; they evaluated whether to provide them. The power flowed one way because she understood the exchange through the old system's logic, where her value was measured in research outputs and they were the gatekeepers who decided what those outputs were worth. But this kept her as a producer.

One of her long-standing industry relationships had felt stuck in the way her career had been feeling stuck. She was always asking. They were always deciding. Until she decided to ask them a different question. Not what do I need from them, but what do they need, and what can I offer them that they can't get anywhere else?

Once she changed her perspective, the answer was obvious. She could offer clinical insights they had no pathway to generate on their own, validation in patient populations they couldn't access and real-world evidence from settings where their technology was actually being used to make decisions about patient care. To replicate what she had created would cost them years and millions, assuming they could even find people with her particular combination of clinical experience and research expertise.

Sarah had been so focused on what she lacked that she'd never catalogued what she had, let alone its value to others. And when she started to frame her conversations with industry partners in this way, as her value to them, the dynamic reversed.

She pitched two project ideas, framed not as requests for support but as opportunities for partnership. Both were structured around the value she could provide and addressed problems they actually had. They were keen on both.

Around the same time, she was assembling a partnership grant. When a potential industry collaborator asked about the scale of contribution she was expecting, she said something she couldn't have said two years earlier – up to $400,000 AUD in cash and in-kind from each partner. They didn't negotiate. They agreed to the full amount.

'A year ago,' she wrote, 'I would never have been thinking in that realm.'

Same researcher. Same science. Same relationships she'd had for years. None of that had changed. What *had* changed was her understanding of her own value, and how to approach her partnerships – that is as a partner, not just as a producer – so she could find those that needed to see her succeed.

Moving from being acted upon to taking action

Each of these moments had led Sarah to this place where her research could align with the real world. But each of these moments required her to act before she was certain. Reordering her conference talk before she knew how the room would respond. Naming her value to an industry partner before she knew whether they'd see it as well. Pitching a number she'd never have dared pitch a year earlier and waiting to see what came back.

There was risk here – risk that she was wrong about what she had, why it mattered or who needed it. The old system had taught her that being wrong was failure and trained her to stay where she was until she was sure. But what Sarah discovered, one uncomfortable conversation at a time, was that being wrong early cost almost nothing, while being wrong late, or never finding out at all, was what kept researchers stranded on the plateau she'd finally been able to escape.

She'd finally been able to move from being acted upon to taking action, from passive producer to active partner, and then momentum was undeniable.

Momentum

Within a year, Sarah was awarded a national award for science communication. The following year, an institute award for leadership and collaboration. Then she earned a promotion, and with that promotion came the mandate to build her own research group, assembling a team around the work she'd spent her career preparing for. Not long after came a major partnership grant, with multiple organisations contributing more than a million dollars to a project she'd shaped and positioned.

Each recognition developed the conditions and built the momentum for the next one. The communication award made her visible to the people who selected the institute honour. The institute honour strengthened her case for promotion. The promotion gave her the platform to attract partnership funding at a scale she'd never attempted before. Sarah's flywheel was well and truly spinning.

So when her faculty conducted an international search for an executive role, they interviewed candidates from multiple countries, reviewed their strategic vision and leadership philosophy, and ultimately hired someone who'd been there all along – Sarah. She had reconnected with why she'd become a researcher in the first place, and by repeatedly building her Strategic Advantage, she had also learned to let others see it too.

Giving yourself permission to act

Sarah's calendar today looks much like it looked five years ago. She's still at the same institute, she's still being published in the same journals and still being invited to speak at the same conferences. But she operates differently now, in ways that would have been invisible to her five years ago.

What changed wasn't her science or her calendar or even her ambition. It was how she saw the exchange. The old system had taught her that good work gets recognised, that if you produce enough quality research the gatekeepers will eventually let you through. She understands now that nobody on that old bridge is coming to help her. But she also knows that another bridge exists to a new world where the crossing depends not on what gatekeepers grant but on her ability to connect her world with theirs, to become a partner with those who need the solutions she offers.

The barrier she'd spent years pushing against hadn't been made of institutional constraints or funding limits or gatekeepers who wouldn't let her through. It had been made by a set of limiting beliefs dictated by how the traditional system said she was allowed to act.

Now she understood that she was allowed to start conversations before she had answers. She was allowed to pitch ideas before they were fully formed. She was allowed to describe her value in terms that the old system didn't value. She was allowed to test, to reach out, to claim her place in conversations she'd assumed she wasn't invited to.

Nobody had told her she wasn't allowed. The old system had simply discouraged it as irrelevant to the metrics it cared to measure. It had pointed her toward one bridge and never thought to acknowledge that another existed.

Today, she starts with problems, not solutions, because problems are what other people recognise. A clinician doesn't care about her mechanism; they care about their patient who can't get answers. A funder doesn't care about her methodology; they care about the gap in care that their patients fall through.

And she sees more in herself than the old system's metrics ever revealed. Those metrics measured publications, citations and grant income, not what she's really worth. Now she knows her value is bigger than those metrics – that it's actually the capability she has to answer questions that only someone with her particular combination of clinical experience and research expertise can answer. And once she saw that clearly, everything else followed.

Not an outlier – a pattern

Sarah's path is one of many. Each of the researchers you met in Part II are proof that the pattern holds across very different journeys, and proof of what becomes possible when Strategic Advantage starts to compound.

Look at what Marcus actually has now. Not just the mechanism, because he'd had that for years. Not just the publications, because he'd had hundreds of those. What Marcus has now is a constellation that didn't exist when he started: investors who believed early and stayed committed, a team built for clinical development, a regulatory pathway already clearing and relationships with people who want to see him succeed.

Marcus's pipeline now doesn't just include the lung application that first attracted believers, but kidney disease and liver disease and neurodegeneration, the applications he'd always envisioned. And each of these are now on their way to creating patient solutions – not despite the commercial journey but because of it. Over $30 million USD has been raised, and the first-in-human clinical trials are being prepared, all from partnerships the old system never revealed to him. All because he saw the path, and started to pull the levers of his flywheel.

Another researcher with the same scientific insight could, in theory, reach the same destination. But they would need years to build what Marcus has built, and by the time they'd done all of that, assuming they could, Marcus would have already reached the patients he wanted to help.

But here's what the numbers don't capture. Marcus isn't the same researcher who sat in his office telling me 'they don't get it'. He still cares about the science with the same intensity he always did. He still traces the inflammatory pathway with the precision of someone who has spent two decades immersed in it. But the man who once believed that the commercialisation office would block any approach to outside Capital, who was certain that venture investors wanted sure things and would never look at science early, who had internalised every assumption the traditional system had taught him about what was and wasn't possible – that researcher doesn't exist anymore.

In his place is someone who reads forces he couldn't see before. A researcher who understands that Capital flows toward demonstrated value rather than waiting to be extracted through applications, who knows that the world was looking for what he'd built all along and who understands that the only thing that stood between his science and the people who could carry it forward had been assumptions that he'd treated like facts and carried for so long.

What James and Ben built, once they could tell the difference between interest and intent, was a company that spun out in 2023, and by late 2024 had raised over $10 million AUD from backers who could see what two years of directionless advice had obscured. They have a team now, including Anna, who brought the operational capability neither of them could have provided alone, and a foundation tested through myriad reality checks that revealed what their science could become.

But what matters most isn't their venture, because ventures are just structures and structures serve the mission rather than being the mission itself. What matters is what the process revealed. James, who once couldn't answer 'who actually wants this?' now understands Demand as a force to be read rather than assumed. And Ben, who had always leaned toward the boundary between science and commerce, discovered that his instinct for translation wasn't a deviation from research but a different expression of the same impulse that had drawn him to science in the first place.

Kusal doesn't check crashed computers anymore. The platform he was building when we met him in Chapter 5, the one he could feel taking shape but couldn't yet see the full outline of, has become a dedicated centre funded by millions in philanthropic Capital that arrived not because he chased it but because the people who controlled it could finally see what he was building. He earned a promotion to associate professor and has been recognised as one of his country's leading innovators. The support that followed came from places that would have seemed fantastical during his early days working from the corner of someone else's laboratory. It came from national research councils, international defence agencies, cancer organisations and a growing network of hospital and philanthropic partners whose funding he once would have spent half his year applying for.

But the deeper change, the one no resume captures, is in how Kusal sees himself. He's no longer a researcher running disconnected projects and scrambling for survival. He understands now that the collaborative infrastructure he assembled from almost nothing was itself the Product the world had been searching for, and that Capital had been there all along, waiting for someone to show Demand what it was looking at.

For Sadhika, her clinical trial is running now, across a national network, generating the evidence that will determine whether the technology can scale to every clinic in the country. The company that first opened its database has deepened its commitment to co-invest-ment. Other Capital soon followed because the fellowship signalled Supply credibility, the industry partnership signalled authentic Demand, and because funders talk to each other, conviction backed by proof compounded faster than anyone involved quite expected.

But what matters more than the trial or the funding is this: Sadhika no longer carries untested assumptions the way she once carried the belief that her camera was her Product. She tests everything now, holds her beliefs firmly enough to act on them and loosely enough to abandon them when a clinic visit or a conversation reveals that what she thought she knew was only partially true. The LARC Loop has become how she thinks, not something she does. Each cycle doesn't just reduce uncertainty. It builds the relationships and the clinical insight and the earned trust of an industry partner willing to put their own resources at risk, the kind of things you can't reverse-engineer from the outside.

None of these researchers have reached their final destination, and none of them expect to anytime soon. All the while they are enjoy-ing the journey and the view from the new bridge, their Strategic Advantage keeps building. What's changed is that they're no longer passive passengers on someone else's journey, waiting for gatekeep-ers to decide their fate. They are partners in the process now, not because someone granted it, but because they claimed it through pulling the right levers, testing their assumptions and taking action towards certainty.

And they are not the exceptions. They are examples of something much larger.

Chapter 10

THE RETURN

In the Introduction, I described a moment in 2015 when two worldviews collided in a public, packed auditorium, with a pilot programme I was trying to launch and a senior researcher who stood up and told the room that these people should be prioritising papers and publications. In that moment, I was certain everything we were working toward was about to collapse. But then, against every expectation, a queue formed of researchers wanting to take the journey.

I never told you what happened to the researchers who stepped forward to join us on the pilot programme that day.

From sand to success

One team that queued up to talk to me after that conference came with a jar of what looked like sand.

It wasn't sand. It was a metal-organic framework — an engineered material with a structure so porous that a single gram contained the internal surface area of a football field. The physics of it still strikes me as slightly magical. You can pack a cylinder with this powder and it can hold 40 times more gas than if the cylinder were empty. Not

40% more. 40 times more. The material creates so many binding sites to hold gas molecules that you can store what would normally require a tank the size of a room in something you could fit in a backpack. You could even design this sand to suit different gases.

What they had was good. The science was validated, the publications were there and the grants had been awarded. And as far as the traditional system was concerned, the work was complete. All that was required was for industry to come knocking. So they put up their shingle, fully expecting people to beat a path to their door. But, as you can likely surmise by now, no one had come.

The researchers had been thinking about industrial applications. These are the kind of use cases that typically appear in journal articles and investor presentations, things like hydrogen storage, carbon capture and transportation of natural gas. They're important applications, certainly, but diffuse. No one was urgently knocking on laboratory doors demanding better hydrogen storage solutions. It wasn't specific enough.

So we asked a different question: who, right now, is searching for a way to hold more of a specific gas in less space?

What they found when they started looking surprised them. Defence agencies were actively searching, but not for gas storage in the way the team had been thinking about it. They were looking for better respiratory protection for their soldiers in the field.

It turned out that the canisters in gas masks hadn't been meaningfully improved since the First World War. A century of technological advancement, and soldiers were still breathing through filters designed when cavalry was a real tactical strategy. But the material this team had developed could change that equation entirely. With 40 times the storage capacity, it also offered 40 times the filtration capacity, and all with the same volume. This meant they could even design the sand to offer advanced protection against new threats the old technology couldn't even touch. And all in the same size package, with little impact on weight.

What followed was a multi-million-dollar defence contract and new manufacturing partnerships that are now building equipment to protect soldiers – and also firefighters, miners and industrial workers. The researchers hadn't changed their Supply, but once they'd looked at their Product differently, they'd discovered that Demand had been there all along, looking for them.

Putting solar cells in space

Another team that joined us had figured out how to print solar cells the way newspapers print pages.

Think about what that means for a moment. Silicon solar panels are manufactured in factories that cost billions of dollars to build, through processes involving high temperatures and toxic chemicals and equipment that takes years to install. This team could print photovoltaic material onto thin plastic film using industrial printing processes that already existed, at room temperature, in rolls that could theoretically be produced by the kilometre.

The efficiency numbers weren't competitive with silicon yet. In a head-to-head comparison of peak output, traditional panels still won.

And the old system's approach had kept them in the laboratory for years, chasing incremental efficiency improvements that would look impressive in publications while the actual technology sat waiting.

The programme pushed them toward a different question: who needs solar cells that are flexible rather than rigid, light rather than heavy and functional even if efficiency isn't the deciding factor?

The answer, once they started looking, was obvious. Not every surface can bear the weight of silicon panels. Not every application needs peak efficiency. A farmer who wants to power an irrigation sensor in a remote field doesn't need the same output as a suburban rooftop installation. A disaster response team that needs portable power doesn't care about 20-year warranties. A building with a curved facade that couldn't support conventional panels might welcome something flexible that could be applied like almost-invisible wallpaper.

Once they understood what game they were actually playing, the old rules didn't apply. Within a few years they did something that highlights the distance between laboratory validation and real-world demonstration: they launched their solar cells into space on a commercial satellite. Their printed power generators were orbiting the earth while the old system was still asking whether the efficiency numbers were good enough.

The traditional path would have kept them refining, assuming they weren't ready. The path they chose instead put them into orbit.

Printing implants on demand

A third project reframed its partnership around a question that sounds simple until you think about it carefully: what if every medical implant could be designed for exactly one patient?

Each human body is unique, and your bones are not the same shape as mine. But when a surgeon needs to replace a hip or reconstruct a jaw or repair a skull after trauma, they're stuck working with implants manufactured in standard sizes, which means they're always compromising. The best they can hope to achieve is close enough, good enough, the nearest approximation of what the patient actually needs.

This team believed they had a better answer. So they connected a research facility's 3D printing capability with a medical device company founded by a surgeon who understood exactly how frustrating those compromises could be. The question they asked wasn't what can we print, but who needs something they can't get any other way?

The answer came faster than anyone expected. A patient was facing amputation because cancer had invaded his heel bone. The standard options were removing the foot or attempting a reconstruction with standardised implants and would likely result in amputation anyway. Neither option was good.

Two weeks from the initial phone call to this team, that patient received a custom-printed titanium heel bone, the first of its kind to be used in surgery anywhere in the world. This was an implant designed from his own scans, manufactured to match his exact anatomy, printed in metal and implanted before the disease could spread further. He kept his foot.

Next, a cancer patient on another continent received a custom-printed sternum and rib cage, another world first. This led to more patients, more procedures and more surgeons around the world who discovered that the compromises they'd been having to make and asking their patients to accept were no longer necessary. They now have a better solution.

The researchers who formed that partnership with industry didn't leave their research behind in order to create impact. They just asked a different question about who needed what only they could build.

Translation isn't always about success

But not every project succeeded, and the failures mattered as much as the successes.

One team had been working on display technology, a way to achieve deeper and more saturated colours on screens using organic LED technology. The science was real and the achievement was genuine – a blue that was, as they put it with pride, more blue than any blue on the Market. They had pushed the boundary of what was technically possible.

The programme encouraged them to get in front of people who understood the display industry. What they learned was uncomfortable. The Market didn't care about that particular metric, the blue was already good enough. The improvement they'd achieved, though impressive in the laboratory, didn't solve a problem anyone else considered to be a problem.

Consumers weren't choosing televisions based on whether the blue was blue enough. And since consumers didn't care, neither did the manufacturers. Industry just wasn't searching for what they were building, and so the project closed.

The researchers were angry at first. They had done years of excellent work, and they were being told it didn't matter. But as the implications settled, anger gave way to relief. They had been heading down a path that would have consumed even more years and still led nowhere. The programme had shown them the dead end before they'd walked too far through their careers to turn back.

Later, they thanked me, as the closure freed them to work on something better. They went on to lead research programmes and significant national research infrastructure that connected multiple projects with problems people actually needed solving. They found more meaning than the display project had ever delivered.

Translation isn't always about success. Sometimes it's about discovering early that something won't work, before years of your career and life disappear into something the world doesn't value. That discovery is its own kind of progress, initially painful but far less painful than the alternative.

A colliding worldview

And what of the senior researcher who stood up that day, who doubled down on the old system's approach? Well, she's certainly not the villain in this story, and her career continued upward. She rose to the executive level in her institute, and in 2021 she was appointed as

Australia's Chief Scientist. In this role, she reached a position where she could see the entire research landscape – not just the piece visible from her own laboratory, but the whole architecture. She could map the flows of funding and talent and ideas that connected basic research to national prosperity. She could see the gaps where promising science fell through and the distance between what researchers produced and what reached the world.

From that height, years after our collision in the auditorium, she said something publicly that caught my attention: 'I knew my research was a potential game changer. But back then, commercialisation wasn't the mindset. The job of a scientist was to publish a paper. This was the end point.'

And later: 'There is no shortage of excellent research. Our discovery research must continue. But let's be frank, our research is not being translated into new Products and innovations nearly as often as it could be.'

The researcher who had told a room full of early-career scientists to prioritise publications was now leading the country in this new dialogue, expanding on the idea that something had gone wrong on the pathway between discovery and deployment. She was building the narrative at a national level that excellent research was not enough, that the bridge that had supported her career was not able to carry what was now being asked of it.

I want to be careful about what this represents, because the framing matters.

She wasn't wrong in 2015 about the world as it was. Publications had mattered (they do matter). The old bridge had worked, for a time, for those who could get across it. Her advice had been sound for the system as it once was. What happened in the years between that auditorium and those public statements wasn't that she changed her mind. It was that from her new vantage point she could see what was happening at the edges of that old world: a second bridge, built by different stakeholders with different incentives, operating on different logic entirely.

In 2015, we were looking at the same problem from different positions. She stood deep inside the old system, protecting what had worked for decades, pointing researchers toward the only bridge she knew. I stood at the edge of something new, on a bridge that was still being built.

By the time she gave those talks, there was no longer a debate about whether the shift was real. The new bridge was working. The funders had built it and their support and Capital was waiting. The question was no longer whether researchers should engage with translation but how to find their path, on their terms. We were two people whose views converged because she rose high enough to survey the territory while I was on the ground building it.

And remember the managing partner of the venture capital firm who had told me, years earlier, that educated researchers made deals difficult? He wasn't wrong about the dynamics he'd observed. Researchers who understood their science deeply but couldn't see how it fit into the larger picture, who waited for the world to recognise their excellence rather than building the Strategic Advantage that would make recognition inevitable, were the norm. Those dynamics were real.

But he was describing a world where the gatekeepers who controlled who could traverse that old bridge benefited from opacity and where the only path across the valley required surrendering control to people whose incentives didn't align with yours.

Marcus, Kusal, James, Ben, Sadhika and Sarah had found their paths to the new bridge. And in doing so, they had become precisely the kind of researcher that this managing partner had never encountered. These were researchers who could sit across a table and articulate not just what their science could do but why anyone should back them to do it, who understood the forces at play and the levers they could pull and who saw the value and Strategic Advantage they had built that no competitor could easily match. And not because they'd abandoned research, but because they'd learned to connect it with the world that needed it. They were now informed partners ready and able to drive its translation.

The senior researcher and the managing partner had both been right about the old system. They had simply been wrong about what was possible when researchers stepped into a different system altogether.

A pattern of translation

That auditorium in 2015 was the beginning of something I couldn't have predicted.

The pilot programme those researchers joined became the foundation for national programmes that continue to this day. The approach we tested in that first cohort – the questions we learned to ask, the patterns we discovered about what made research move – all of it was sharpened through iteration and through working with researcher after researcher facing variations of the same challenge.

I have since had the privilege to work closely with hundreds of motivated researchers over the years. Not a handful of exceptional individuals who happened to figure it out, but hundreds, across disciplines and institutions and career stages, each one testing whether a pathway existed for them and finding that it did.

I have personally seen hundreds of millions of dollars flowing toward the work of these researchers, enabling development the traditional system could never have supported. They learned to see funding sources the traditional system never mentioned and built positions that attracted Capital in all its forms.

The support they secured hasn't come from the traditional grant pools they'd been trained to chase. It's flowing across the bridge this book has been bringing into view. It's from disease foundations, venture philanthropists and mission-driven funders who have built infrastructure to catch what the old system dropped. It's from new industry partnerships and progressive venture capitalists, and even new government programmes designed for translation rather than discovery. In contrast to the old world, Capital has been increasing year by year, just waiting for researchers who are ready to cross to this new world.

Myriad projects are now in flight, through companies, partnerships, programmes and platforms. This is science moving toward the world rather than sitting in labs or documented in journals waiting to be noticed. Some will succeed. And some will fail. But all of them represent researchers who decided to test whether another path existed and discovered that it did.

This isn't a handful of stories, and the shift isn't theoretical – it's a pattern, proven at scale, over a decade and counting. The funders have already moved. The infrastructure already exists. The Capital is waiting.

Remember the foundations gathered in that LA conference room, frustrated that they couldn't find researchers ready to meet them? This is what they were – and are still – looking for. Not good science, because they'd seen plenty of good science disappear into the void. Not promise, because promise is cheap and potential is abundant. They were looking for researchers whose flywheels had started to spin. Researchers whose centre of gravity had shifted from Supply to Demand, from building in isolation to building within the world. Researchers who still engaged with grants and infrastructure when those things were worth having, but who were no longer waiting for permission to test in the real world and to start crossing that bridge.

Of the researchers who stepped forward in that auditorium in 2015, some are now working on next-generation protection systems, some have opened manufacturing facilities, some are exporting medical devices globally and some discovered early that their projects weren't going to work and redirected their expertise toward something better.

Marcus and Kusal and James and Ben and Sadhika and Sarah – they are what those funders have been searching for. Researchers who learned to read the forces, who built demonstrated value and whose flywheels are spinning. Researchers who found the new bridge and are walking it.

The next step is yours

At the beginning of this book, I described the researchers who had made this crossing in general terms, because you had yet to meet them. I told you they weren't more entrepreneurial by nature, weren't more willing to compromise and weren't different in kind from anyone else. I asked you to take that on faith. You don't have to take it on faith anymore.

You've seen Marcus sit in a laboratory for 20 years with validated science that couldn't find a path. And then you saw him discover that the path had been there all along, just not where the old system had taught him to look.

You've watched James and Ben drift for two years, pulled in every direction by people who could see pieces of the opportunity but not the whole. And then you watched them build a foundation precise enough that serious Capital recognised what they had.

You met Sadhika as she walked into a dermatology clinic convinced her camera was the Product. And you saw her walking out, understanding that the algorithm was the Product, a revision that took a single afternoon and changed the trajectory of everything that followed.

You've sat with Sarah as she reordered a conference talk. And you were with her when she discovered that the audience she'd been speaking to for years had been waiting to hear something she'd always known but never thought to share.

None of these researchers had a map. All of them had the same training you had, operated under the same assumptions and within the same system that kept pointing them toward that bridge that couldn't carry them and their work. But what they did differently was move before they were certain, test before they were ready and discover through action what no amount of preparation could ever have revealed.

Every one of them started where you are now. Every one of them carried the same assumptions the old system installs. And every one of them found their way to a bridge the old system never showed them. The next step is yours.

Chapter 11

THE NEXT GENERATION

Every researcher in this book had to do something that, because of their training, felt wrong before they discovered it was right. Some were pushed by circumstances, funding that dried up or paths that closed. Some were pulled by opportunity, something emerging they wanted to be part of. All of them felt the fears and doubts that any researcher feels when they contemplate stepping outside the path they were trained for. They stepped anyway.

But I've noticed something over the past decade. The crossing is getting easier for those who come after. The bridge has been built. The researchers who came before did this for those who are coming after.

One of the last researchers I worked with while finishing this book is a PhD student named Beth.

She was still completing her doctorate when she entered our programme, researching a disease where patients cycle through flares and remissions, never quite sure when the next attack is coming. The science was promising. The question she couldn't answer from inside the laboratory, however, was whether anyone outside it would ever care enough to carry the work forward.

What struck me about Beth wasn't the quality of her science, though the quality was there. It was that she asked me directly whether her project might be worthless. Not quite in those words, but she came looking for answers to the hard questions – the ones that could kill the project – because she needed to know the answers before she spent years on something that might never matter.

She sought out the critique rather than avoid it, driven to test her assumptions against reality rather than protect them from it. The diagnostic test she thought was needed wasn't the Product the world needed, and she discovered this the same way every researcher in this book did: by testing. So she adjusted, not abandoning her science but redirecting it toward a problem that only became visible once she stopped protecting her assumptions from it.

By the time she finished the programme, she had started conversations with investors, engaged health economists, connected with manufacturers and learned how to present her work in language that made sense well beyond academia. All while still completing her PhD. And all before the conditioning of the traditional system had fully set in.

Because of this, Beth didn't have to unlearn much. She encountered the new bridge before the old one had calcified into the only way she saw. The questions that felt revolutionary to Sarah or Marcus or Ben and James, felt obvious to Beth, because she'd been asking them from the beginning. And because no one had told her that the old bridge was the only path.

Beth's not alone. Across the programmes I run, I'm feeling the shift. More researchers are arriving who already know the old bridge can't carry them. They don't need to be convinced there's another way forward. They come looking for the tools and tactics to find their way to it. This next generation is forming around a different centre of gravity, and the researchers they go on to mentor will hopefully see both bridges and give themselves permission to cross either, or both, as the need requires.

The old system sustained itself through generations of researchers training the next generation in patterns that had worked for them, pointing each cohort toward the same bridge their mentors had crossed. The new system is building itself differently, one researcher at a time, each crossing making it easier for the one behind to see that another bridge exists.

You have to make the move

The old system that trained you assumed the world would meet you halfway. It won't. The bridge it pointed you toward is overcrowded and failing, controlled by gatekeepers who benefit from the scarcity. Waiting for it to start working again is no strategy at all.

But people are standing on the other side of the valley, people who spent decades building a different bridge because they got tired of watching excellent research die in the crossing. People whose success depends on yours. People who are actively searching for researchers ready to meet them.

This book was written to introduce you to each other – you and the funders who are searching for you, the partners who need what you've built and the patients who are waiting for solutions that research has made possible but the old system has yet to deliver.

Now the introduction is made. What happens next is yours to decide.

CODA

The flight home from LA in 2019 was long enough to think.

The foundation leaders in that Beverly Hills conference room had been searching for researchers who didn't know they were being searched for. They had built infrastructure, deployed Capital and constructed a bridge across a valley that had been swallowing so much promising science for decades.

They were frustrated, genuinely frustrated, because the researchers they needed couldn't see them, and they couldn't see the researchers, and the distance between the two wasn't geography or gatekeeping but something harder to reconcile: a frame of mind that kept both sides invisible to each other. So the foundation leaders kept waiting and the researchers remained queuing for the old bridge, all while a new one stood waiting.

I thought about the researchers I'd worked with over the years who had made the crossing, and the ones who hadn't. I thought about the ones who had learned about the new bridge but were still deciding. I thought about what separated them, and the answer that kept surfacing was less dramatic than I'd expected.

It wasn't talent, because I'd watched brilliant researchers stay stuck and ordinary ones break through.

It wasn't entrepreneurial instinct, whatever that means, because most of the researchers who crossed would never have described themselves that way, and most of them still wouldn't.

It wasn't even circumstance, though circumstance played its part, because researchers facing identical constraints made different choices about what those constraints allowed.

What separated them, as far as I could tell, was permission.

This was permission to see themselves differently than the old system had taught them to see themselves. Permission to question assumptions so deep they'd forgotten they were assumptions at all. Permission to reach out for feedback before they felt ready. And permission to claim a place in conversations they'd assumed they weren't invited to.

The permission was the hard part. Everything else could be learned.

Somewhere over the Pacific, the book I hadn't known I was going to write took shape.

You have permission

As you've got this far, there's one thing I know can't be undone – once you have seen the new way, you can't unsee it. Once you understand that another way across the valley exists, that the funders who built it are waiting and that the crossing depends on what you build rather than what gatekeepers grant, you can't return to the comfortable blindness of before. The researchers who cross become something different in the process. They become the kind of researcher our world needs right now.

I don't know what constraints you're facing. I don't know your field, your institution, your circumstances or the particular shape of whatever's keeping you where you are. I don't know whether you picked up this book because something already shifted or because you're hoping something might.

What I know is that a jar of sand that holds 40 times more gas than an empty cylinder is now being built into equipment that will save people's lives. That solar cells printed like newspapers are orbiting the earth. That a man kept his foot instead of losing it to the old system. That a researcher who couldn't get anyone to care about her work is now shaping how an entire faculty thinks about research. That a PhD student's algorithm is running clinical trials to better detect cancer. That Capital is now flowing toward a mechanism that sat in journals for years but could go on to treat hundreds of millions of patients. All because someone asked the right question about who needed what to exist.

None of that was inevitable. All of it required someone to move first. To start their flywheel turning.

Everything you need to make the crossing is already in your hands. And now you have permission. The first step is how your *breakthrough* begins.

BEHIND THE BOOK

If you've read this far, you may have noticed that this book doesn't feel quite like other books in this space. That was deliberate.

Most business books, and self-help books for that matter, offer a linear process as a panacea for common problems. Read the chapters in order, follow the steps and apply the framework. It's a compelling promise, but one that rarely survives contact with reality, because reality doesn't hold still while you work through the steps. Every action produces a reaction. Every conversation shifts what you know. And because of that, the terrain you're navigating today won't be the terrain you're navigating next month.

A linear process can't help you with that. What helps is the ability to read the landscape as it changes, and to know what you're looking at when you do.

That's what this book was designed to build.

The frameworks are real. The methodology works. But I didn't lead with that IP, or organise the book around it, because 15 years of working alongside researchers taught me something that changed how I design everything I do: knowledge doesn't produce behaviour change.

Researchers already have knowledge. As a researcher, you already know you should talk to industry. You already know translation matters. You already sensed (and now you know) the system is broken. What you're lacking isn't information. It's the insight that allows you to see your own situation clearly enough to act, and to keep acting as the situation shifts around you.

The approach I've taken in this book draws on a body of research developed through military and industrial training programmes where the stakes were high enough that the training had to work. Two theories in particular have shaped how I design my programmes, my coaching and this book.

The first, Cognitive Flexibility Theory, says that expertise in complex domains is built by encountering the same principles across many different cases. The second, Cognitive Transformation Theory, developed by the psychologist Gary Klein, says the most powerful thing you can do for a learner is not instruct but transform by dismantling the mental models holding them back and replacing them with the causal understanding that experienced practitioners actually use.

Klein is right. I've worked alongside hundreds of researchers at every career stage, from PhD students to institute directors, and the most important breakthrough is never the methodology. It's the moment when your understanding of the system shifts. When you stop seeing the traditional pathway as the only pathway. When you start noticing opportunities that had been visible all along but invisible because your mental model had no place for those opportunities to land. Once that shift happens, behaviour changes. Not because anyone told you what to do, but because you can suddenly see what to do, and the frameworks that had felt abstract become tools you reach for instinctively.

The stories came first, deliberately, so that by the time the frameworks arrived you'd already begun to recognise what they were describing. They emerged from the stories before the frameworks gave them shape, so that by the time you encountered them formally, they felt like language for something you'd already begun to understand.

Which is how it should feel. If you start to cross the new bridge, it won't be because you followed my instruction most faithfully. It will be because your understanding shifted, because you started seeing forces and levers and loops where you previously saw only obstacles or luck. That shift tends to be permanent. It changes not just what you know but what you see, think and do, and how you interpret everything that happens to you afterwards.

If you find yourself, in the weeks and months ahead, noticing things you didn't notice before, seeing Demand where you previously saw indifference or recognising the shape of a learning loop in a conversation with a clinician or a funder or a patient advocate, then the book did what it was designed to do.

Thank you for reading. It has been a privilege to share this work with you.

If you'd like to go further than a book can take you, I'd welcome the conversation: hello@dougaledwards.com.